Andrew Wilson

Solar System Log

P9-AGU-183

JANE'S

Copyright © Andrew Wilson 1987

First published in the United Kingdom in 1987
This edition published in 1987 by
Jane's Publishing Company Limited
238 City Road, London EC1 2PU

ISBN 0 7106 0444 0

Distributed in the Philippines and the USA and its dependencies by
Jane's Publishing Inc,
115 5th Avenue,
New York, NY 10003

Computer typesetting by D.P. Media Limited,
Hitchin, Hertfordshire

Printed in the United Kingdom by
Biddles Limited, Guildford, Surrey

Acknowledgements

My thanks are due to Phil Clark, Nicholas Johnson and David
Woods for their extensive assistance with the Soviet sections;
Phil Clark especially has been an untiring source of infor-
mation and advice. I also thank them for reviewing the early
drafts, although any errors remaining must be regarded as the
responsibility of the author. Extensive time and assistance
have been freely given by a wide range of individuals involved
in the world's space activities, including Dr Roger Craig
(NASA Ames Research), Dr Robert Farquhar (NASA Goddard
Space Flight Centre), Judith Herb (Martin Marietta), Prof Kunio
Hirao (ISAS), Dr Norman Longdon (ESA), Dr Yasunori
Matogawa (ISAS), Prof Tony McDonnell (University of Kent),
Stephen Tellier (Lunar & Planetary Institute), Dr Ewen Whitaker
(University of Arizona Lunar & Planetary Laboratory), Alan
Wood (JPL), NASA, the Novosti press agency, Ralph Gibbons,
Rex Hall, Reginald Turnill and Tim Furniss.
  To all of these I extend my appreciation.

Andrew Wilson
London
September 1986

2

Foreword

Jet Propulsion Laboratory director Dr Lew Allen.

For more than a quarter-century the United States has been systematically exploring our solar system. Men have landed on the Moon, studied its surface and returned samples of lunar material for laboratory study. Spacecraft have visited each of the planets as far out from the Sun as Uranus, and the indomitable Voyager 2 will encounter Neptune in 1989. The Soviet Union has also conducted a programme that has included unmanned exploration of the Moon and, in recent years, a very successful investigation of Venus.

Recently the study of the solar system has become international, with an armada of spacecraft flying by Halley's Comet to study the characteristics of that historic visitor from the deepest regions of the Sun's influence. The flyby fleet consisted of two Soviet, two Japanese and one European spacecraft. In the future there will be more challenging missions, more international participation and more surprising and important discoveries.

The primary scientific goal of planetary exploration is the determination of the origin, evolution and present state of the solar system. Investigators seek also to understand the Earth through comparative planetary studies, and the relationship between the chemical and physical evolution of the solar system and the appearance of life. Some see the survey of resources available near the Earth for possible future human use as a fourth rationale for space exploration.

The exploratory missions to date have contributed greatly to the achievement of these goals. A wealth of information has permitted a growing understanding of the processes that determined the nature of the planets and their atmospheres, moons and rings. We have gained a glimmer of understanding of why the Earth is unique and how it happened that only here in our solar system have conditions developed conducive to the evolution of life. Yet these marvellous voyages of scientific discovery have been only a reconnaissance of the major features of the varied and complex elements of the solar system. The next 25 years will see missions that may include rendezvous with and close study of comets; close flyby of several asteroids; rendezvous with and landing on Phobos; more detailed observation of Mars, Venus and the Moon; a probe deep into the atmosphere of Jupiter; very close observation of the Galilean moons; a probe into the atmosphere of Titan; a rendezvous with the rings of Saturn; and the beginnings of a programme to return samples from several planets, particularly Mars.

Although the United States, and presumably the other spacefaring nations, have formulated their planetary exploration programmes on the basis of scientific objectives, they also have other purposes. National prestige is certainly an important factor. The public associates a country's technical prowess with its space accomplishments, and its greatness with its achievements in this inspired and visionary field. Each nation has recognised that the technical challenge and the lofty purposes of space exploration attract the best minds, and that the resulting technological progress is likely to lead to economic benefits. It is also true that the dramatic and often beautiful results from the interplanetary voyages have inspired and motivated the public in beneficial ways. Many scientific textbooks use pictures and other material from the planetary programmes to illustrate fundamental principles. There is no doubt that the human spirit has been illuminated by the wonders of space travel, and that inquiring young people have been given a broader view of life.

What surprises await in the future? The encounter with Uranus proved that the unexpected is the commonplace when exploring new worlds. Are there other planetary systems about other stars? The odds seem to favour it, and strong supporting clues have recently been found. A remarkable observation of Beta Pictoris by an Earth-based telescope shows an extended disc of material about the star, just as we think our solar system would have appeared early in its evolution. Surely our understanding will increase and perhaps someday we will be able to answer that fascinating question: are there really other worlds out there?

Dr LEW ALLEN
Director, Jet Propulsion Laboratory
California

*"When the solar system is conquered,
man will have three dimensions"*

K. E. Tsiolkovsky, 1929

Introduction

The Moon and planets have long fascinated mankind, the Moon as our nearest celestial neighbour and the planets as small discs of light shifting nightly among the zodiacal constellations. But it is only since the dawn of the Space Age in 1957 that we have had the means to send robot probes into the depths of the solar system to transform these objects into three-dimensional worlds. Human eyes have now seen the hidden side of the Moon, the battered face of Mercury, the sulphur-spewing volcanoes of Jupiter's moon Io, the intricate ring system surrounding Saturn and the diverse features of Uranus' Miranda. Mars and Venus have received dozens of metallic visitors, March 1986 at last saw a comet being investigated in detail, and three probes are at this moment breaking

Below: Spaceflight as depicted in de Bergerac's The Voyage to the Moon *(1649).*

Right: Jules Verne's classic science-fiction works inspired this artist's impression of a coal-powered lunar rocket.

free of the Sun's gravitational grasp to penetrate interstellar space for the first time.

The launching of unmanned probes is a vitally important phase in the investigation of the solar system in preparation for colonisation. For example, before lunar bases can be established over the next few decades, the Moon must be surveyed by a polar-orbiting satellite equipped to study the chemical and mineralogical make-up of the entire globe. There is a possibility that the eternally shadowed polar regions harbour water ice that could be processed into rocket propellants. The Soviets have already announced such a lunar surveyor for 1991.

History

The possibility of interplanetary travel has stimulated writers' imaginations for centuries. Cyrano de Bergerac's *The Voyage to the Moon* (1649) described the author's fantastic attempts at a lunar voyage, including a spring-powered launching from a hill. Jules Verne's science fiction classics *From the Earth to the*

Below: Konstantin Tsiolkovsky, schoolteacher and pioneer of astronautical theory.

Below right: Robert Goddard, pictured here in 1924, launched the world's first liquid-propellant rocket.

Bottom: US lunar and planetary launchers. From left: Thor-Able (Pioneer 2 launch), Juno II (Pioneer 3), Atlas-Able (Atlas-Able 5), Atlas-Agena D (Lunar Orbiter 5), Atlas-Centaur (Surveyor 3) and Titan-Centaur (Viking 1).

Moon (1865) and *Round the Moon* (1870) adopted technology which was more advanced – a large cannon – but equally unworkable. It was not until the works of Russian Konstantin Tsiolkovsky (1856–1935), American Robert Goddard (1882–1945) and others that the concept of interplanetary travel was placed on a firm footing. Goddard caused a stir in December 1919 when the Smithsonian Institution in Washington DC published his long, dry scientific paper *A Method of Reaching Extreme Altitudes*, originally written in 1914. Towards the end he mentioned that a means of proving a

The earliest true planetary flights were simple flyby or impact missions, the best that could be managed with technology and the sizes of launcher and upper stage then available. The Americans began by lobbing a few kilos towards the Moon with their Juno II and Thor-Able vehicles. The more powerful Atlas-Able combination was then used in an effort to establish a lunar orbiter, but all three attempts failed. Atlas-Agena launched many of the US flights in the 1960s before the large cryogenic liquid hydrogen/liquid oxygen Centaur was introduced and used well into the 1970s. The most powerful of all US unmanned expendable launchers, Titan 3E-Centaur, saw service with Voyager and Viking in the mid-1970s. The Space Shuttle, with Centaur or other stages, was intended to handle the few US planetary missions planned for the next decade, but the loss of *Challenger* in January 1986 prompted a thorough review of the advisability of carrying the liquid upper stage inside a manned vehicle, and the concept was expensively dropped in June 1986.

The Soviets, on the other hand, have employed a much smaller array of launchers for their deep-space exploits. The A-1 and A-2-e, based on the SS-6 Sapwood ICBM, boosted most of the early Russian probes, with Proton coming on line in 1969–70 for later missions. Europe and Japan have now joined the deep-space club with their Ariane and Mu-3S vehicles respectively.

Left: *The Soviet A-2-e launcher was used for planetary/lunar missions from 1960 to 1972. This picture was released at the time of Venera 7.* **Centre:** *The Proton D-1-e was introduced to the deep-space scene with the Zond lunar craft in 1967 and continues in service. Claimed performance includes 5,700kg to the Moon, 5,300kg to Venus and 4,600kg to Mars. Its next planetary mission will be with the 1988 Phobos probes.* **Right:** *One of the December 1984 Vega launches.*

The missions

Flyby/impact missions provide only brief dwell times close to the target. An orbiter is necessary for global coverage and extended data-gathering, which means that a large rocket motor has to be carried to brake the spacecraft into a closed path around the body.

A soft landing provides an opportunity for the most intimate scrutiny of all, though initially at only a single locale. A more sophisticated soft-lander would perhaps deploy a roving vehicle capable of moving around the surface. Landers usually carry instruments to perform a simple soil analysis – with X-ray fluorescence techniques, for example – but returning a sample to Earth permits a much more detailed examination. Thus far only lunar samples have been made available to terrestrial laboratories, but Mars and cometary material could be available soon after the turn of the century.

rocket had escaped from Earth would be to fire it at the New Moon carrying a load of flash powder to ignite on impact. "The light would then be visible in a powerful telescope," he said, estimating that the yield from less than 3lb (1.4kg) of powder could be detected with a 12in (30.5cm) diameter instrument. He did however warn that the idea ". . . is not of obvious scientific importance." The American press picked up this spectacular concept and ran fanciful headlines typified by "New Rocket Devised by Prof Goddard May Hit Face of the Moon" (*Boston Herald*, January 12, 1920). Goddard even received offers from volunteers to travel aboard his non-existent rocket.

Deep-space missions

	Flyby/impact	Atmospheric	Orbit	Land	Rover	Sample return
Mercury	A					
Venus	A	A	A	A		
Moon	A	—	A	A	A	A
Mars	A	A	A	A		
Jupiter	A	P	P			
Saturn	A					
Uranus	A					
Neptune	P					
Pluto						
Comets	A	A				
Asteroids	P					

A = accomplished. P = firmly planned. Though not yet approved, Mars (rover, sample return), Saturn (orbiter) and comet (orbiter, sample return) missions are all likely to take place over the next 15 years (see Appendix 1).

The Moon has provided a convenient target for mankind's planetary attempts, having been the objective of all the mission types outlined above by 1970. US interest peaked in 1966–67, while the second generation of Soviet landers and orbiters was most active in 1965–66. Following the abandonment of the USSR's manned lunar programme, a second wave of more sophisticated unmanned Soviet craft peaked in 1969–70.

Mars and Venus, the next most accessible bodies, are at the lander stage. The Russians have a long record of Mars failures, in contrast to their Venus successes, but that has not deterred them from a very ambitious Phobos mission in 1988 (see Appendix 1).

No Soviet probes have ventured beyond the three near targets, and only one US vehicle (Mariner 10) has journeyed to Mercury. No further Mercurian missions are being planned, although windows have been identified (Appendix 1).

The four US Pioneer and Voyager probes have all brought the outer planets to the flyby stage – Neptune will be visited by Voyager 2 in August 1989 – and the international Halley missions gathered comets into the fold in early 1986. Jupiter is awaiting its first orbiter and atmospheric probe, the much delayed Galileo. Comet and asteroid rendezvous for long-term investigations or even sample return are likely in the 1990s or soon after.

The entries under each heading in the following table represent the number of successes and failures; eg, 0/4 indicates no successes and four failures for that year. The Soviet totals include the failures listed in the launch chronology table (see page 12) but not the man-related Zonds 4–8. The US totals exclude the Explorers because, although they were lunar orbiters, their studies were not directed at the Moon. The table highlights the fact that the Soviets had to wait until 1967 for their first planetary success (Venera 4) despite numerous attempts at both Mars and Venus. Also remarkable is the fact that no US planetary missions have been launched since 1978: the chronology shows that four Veneras, Giotto and the two Japanese Halley probes have been the only intruders into deep space during 1979–1986.

Soviet success with the first Sputniks led to prompt attempts at the Moon (1958), Venus (1961) and Mars (1960), although many launches failed and remained undisclosed. In April 1959 the Jet Propulsion Laboratory in Pasadena, California, which has since become America's premier centre for deep-space projects, produced an ambitious planetary exploration timetable for the attention of the newly formed National Aeronautics and Space Administration (NASA):

Right: Sergei Korolev (the guiding genius behind early Soviet manned and deep-space successes) seen here with Yuri Gagarin before his untimely death in 1966. Centre right: Georgi Babakin became chief designer for lunar and planetary probes in 1965 and, before his death in 1971, worked on the new Veneras. Far right: V. M. Kovtunenko has been in charge since 1978.

Unmanned Moon, Venus and Mars launches 1958–86

Year	US Moon	US Venus	US Mars	USSR Moon	USSR Venus	USSR Mars	Totals US	Totals USSR
58	0/4	—	—	0/4	—	—	4	4
59	0/2	—	—	2/2	—	—	2	4
60	0/2	—	—	0/2	—	0/2	2	4
61	—	—	—	—	0/2	—	0	2
62	0/3	1/1	—	—	0/3	0/3	5	6
63	—	—	—	0/3	—	—	0	3
64	1/1	—	1/1	0/2	0/2	0/1	4	5
65	2/0	—	—	1/5	0/3	—	2	9
66	3/1	—	—	5/2	—	—	4	7
67	6/1	1/0	—	—	1/1	—	8	2
68	1/0	—	—	1/0	—	—	1	1
69	—	—	2/0	0/5	2/0	0/2	2	9
70	—	—	—	2/1	1/1	—	0	5
71	1/0	—	1/1	1/1	—	0/3	3	5
72	1/0	—	—	1/0	1/1	—	1	3
73	—	1/0	—	1/0	—	1/3	1	5
74	—	—	—	1/1	—	—	0	2
75	—	—	2/0	0/1	2/0	—	3	2
76	—	—	—	1/0	—	—	0	1
77	—	—	—	—	—	—	0	0
78	—	2/0	—	—	2/0	—	2	2
79	—	—	—	—	—	—	0	0
80	—	—	—	—	—	—	0	0
81	—	—	—	—	2/0	—	0	2
82	—	—	—	—	—	—	0	0
83	—	—	—	—	2/0	—	0	2
84	—	—	—	—	2/0	—	0	2
85	—	—	—	—	—	—	0	0
86	—	—	—	—	—	—	0	0

The JPL schedule placed equal emphasis on planetary programmes but NASA headquarters – even in those pre-Apollo days – insisted on a strong bias towards lunar investigations. When Apollo became predominant in NASA from 1961 onwards, interest was focused on unmanned missions that would help the agency to put a crew on the Moon before 1970.

Objective	Achieved by	Soviet equivalent*
Lunar test, Aug 60	Ranger 1, Aug 61	Luna 1
Mars flyby, Oct 60	Mariner 4, Jul 65	Mars 1
Venus flyby, Jan 61	Mariner 2, Dec 62	Venera 1
Lunar rough landing, Jun 61	—	Luna 9
Lunar orbiter, Sep 61	Lunar Orbiter, Aug 66	Luna 10
Venus orbiter, Aug 62	Pioneer Venus, Dec 78	Venera 9
Venus entry, Aug 62	Pioneer Venus, Dec 78	Venera 3
Mars orbiter, Nov 62	Mariner 9, Nov 71	Mars 2
Mars entry, Nov 62	Viking, Jul 76	Mars 2
Lunar orbit/return, Feb 63	—	Zond 5
Lunar soft landing, Jun 63	Surveyor 1, Jun 66	Luna 16
Venus soft landing, Mar 64	—	Venera 7

* Some of the Soviet equivalents failed; others had slightly different objectives to their US counterparts (eg, Zond 5 did not enter lunar orbit).

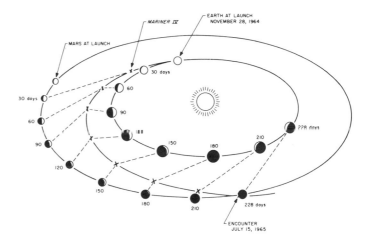

Below: Mariner 4 pursued a typical transfer orbit to Mars.

Getting there

The problems of reaching the Moon were eased by the fact that if a launch failed only another month had to pass before the necessary celestial alignment was repeated. The planets, being more distant, appear in the correct positions to create launch windows much less frequently: every 19 months for Venus, 25 for Mars. The most economical way of reaching Venus is to fire a craft towards the inner solar system so that it reaches its closest point to the Sun (perihelion) at the same time and place as the planet and 180° (half an orbit) away from its starting point. Such a minimum energy transfer orbit allows the mass of the probe to be maximised. Some payload could however be sacrificed to make way for extra upper-stage

Below: The 1962 path of Mariner 2 illustrates a Type I trajectory to Venus.

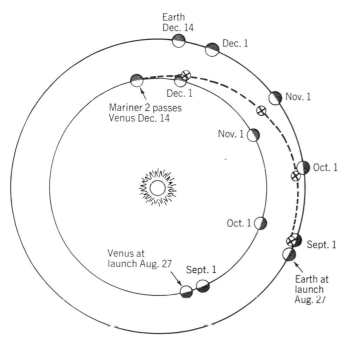

propellant, permitting the probe to travel faster and intercept Venus without having to journey halfway around the Sun. This quicker route, known as a Type I trajectory, shortens the transit time. The Soviets, with their problem of spacecraft unreliability, adopted this method for all of their Venus vehicles until Vega in 1984.

A Type II path sends the probe more than halfway around the Sun, resulting in a long flight time but a lower speed of arrival at the target. NASA did this with its Pioneer Venus Orbiter in 1978 to keep the size of its retromotor to a minimum.

The pattern of Venus launch windows repeats itself every eight years, and because the Earth–Venus paths around the Sun are very circular the energy requirements vary very little every 19 months. Mars, by contrast, has a fairly eccentric orbit and the best windows are 15 years apart; 1971 and 1986 were the most recent.

Trajectory planning has now become fundamental to the design of every deep-space mission. Probes rarely travel directly to their target but swing by other bodies to pick up speed and/or change direction. Mariner 10 was the first to benefit from this technique when Venus was used in February 1974 to direct it towards the one and so far only encounter with Mercury. Voyager 2 has become the prime example: flybys of Jupiter, Saturn and Uranus were all carefully calculated to permit arrival at Neptune in 1989. International Comet Explorer also executed a complex series of swingbys of the Moon to reach Comet Giacobini-Zinner in 1985.

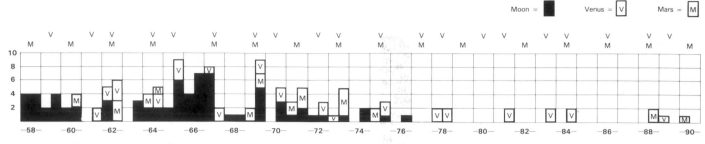

Venus and Mars launch windows

Yearly totals of lunar (black areas), Mars (indicated by "M") and Venus ("V") probes launched by the US (left-hand side of each year) and the USSR (right-hand side). M and V above the columns indicate those years with Mars and/or Venus launch windows. Data taken from tabulation on page 8.

Planet	Date	Exploited by
Mars	Oct 1960	USSR
Venus	Feb 1961	USSR
Venus	Jul/Sep 1962	USSR/US
Mars	Oct/Nov 1962	USSR
Venus	Mar/Apr 1964	USSR
Mars	Nov 1964	USSR/US
Venus	Nov 1965	USSR
Mars	Jan 1967	—
Venus	Jun 1967	USSR/US
Venus	Jan 1969	USSR
Mars	Feb/Mar 1969	USSR/US
Venus	Aug 1970	USSR
Mars	May 1971	USSR/US
Venus	Mar 1972	USSR
Mars	Jul/Aug 1973	USSR
Venus	Nov 1973	US[1]
Venus	Jun 1975	USSR
Mars	Sep/Oct 1975	US[2]
Venus	Jan 1977	—
Mars	Oct/Nov 1977	—
Venus	Aug/Sep 1978	USSR/US[3]
Mars	Nov 1979	—
Venus	Mar/Apr 1980	—
Venus	Oct/Nov 1981	USSR
Mars	Jan 1982	—
Venus	Jun 1983	USSR
Mars	Feb/Mar 1984	—
Venus	Dec 1984	USSR[4]
Mars	May 1986	—
Venus	Aug 1986	—
Venus	Mar/Apr 1988	—
Mars	Jul/Aug 1988	USSR: Phobos
Venus	Oct/Nov 1989	US: Magellan[5]
Mars	Sep/Oct 1990	US: Mars Observer[6]
Venus	Jun 1991	
Mars	Oct/Nov 1992	
Venus	Jan 1993	
Mars	Nov 1994	USSR: Vesta

Type I paths except as follows: [1] + Mercury flyby. [2] Viking used Aug/Sep Type II window. [3] Pioneer Venus Orbiter used May Type Ii window. [4] First use of Type II by USSR. [5] Magellan will use Type II April 1989 window. [6] Mars Observer will use Aug/Sep Type II window unless delayed to 1992.

The payload

Though today's planetary probes carry a larger number of instruments than their predecessors, scientific equipment still accounts for only a small proportion of total spacecraft mass. Giotto's, for example, amounted to only 60kg out of a total launch mass of 960kg.

Instruments can be divided into two broad categories: remote sensing (for flybys and orbiters) and direct sampling (for atmospheric probes and landers). If investigations are at an early stage and the payload space is available, they are selected to provide a broad view of a target. The two Pioneer Venus craft of 1978 exemplify the range of possible studies. The instruments on the orbiter included:

- **Radar mapper** ● Radar pulses returned from the surface were processed to make the first Venus map.
- **Ultra-violet spectrometer** ● Measured the intensity of light at selected UV wavelengths for cloud and haze layer studies. Gases in the upper atmosphere fluoresce characteristically when excited by UV in sunlight.
- **Infra-red radiometer** ● Measured IR/thermal emissions at selected wavelengths to produce vertical atmospheric temperature profiles. Missions now being planned will carry IR spectrometers to the asteroids, the Moon and Mars to produce mineralogical surface maps.
- **Ultra-violet photopolarimeter** ● Returned UV cloud images; polarisation of reflected sunlight depends on the size and density of cloud particles.
- **Mass spectrometers** ● Identified species of atoms in the upper atmosphere. The atmospheric probe also carried one for the lower atmosphere to identify cloud constituents.
- **Electric field detector** ● Measured the electric field components of plasma and radio waves to investigate how the solar wind interacts with the ionosphere.
- **Solar-wind plasma analyser** ● Determined the speed, density, temperature and direction of travel of the solar wind/ionosphere.
- **Electron temperature probe** ● Measured the thermal characteristics of the ionosphere for a study of solar wind heating.
- **Magnetometer** ● Mapped magnetic fields by measuring the electric signal induced in a solenoid-type device. Usually mounted on booms to avoid the craft's own magnetic field, magnetometers can also return data from interplanetary space which are useful in mapping the Sun's own field.

The five atmospheric probes included:

- **Gas chromatograph** ● One entry probe sampled the lower atmosphere three times during descent to determine its content by measuring the characteristic times that gases took to flow along chromatograph columns.
- **Nephelometer** ● Light shining through a window was backscattered by cloud and haze particles and detected after passing through a second window.

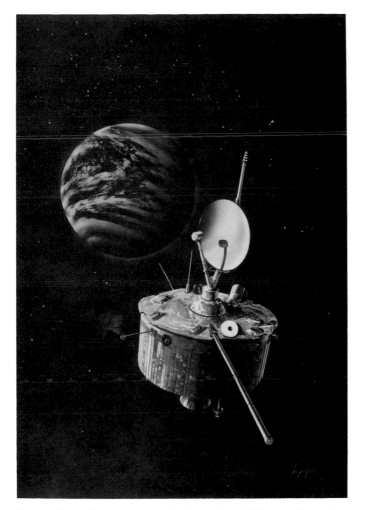

Pioneer Venus Orbiter, launched in December 1978 and still returning data, carries a wide range of instruments for probing the planet (see text).

● **Cloud particle size spectrometer** ● Fired a laser to a prism mounted 15cm outside the probe and detected the return beam to measure 1–500μm-sized particles.
● **Temperature and pressure sensors**.

Tracking an orbiter can reveal variations in the planet's gravitational field (this is how Lunar Orbiter found the Moon's under-surface mass concentrations, or mascons), and monitoring the variation of radio signals as a spacecraft slips behind a planet as viewed from Earth produces information on atmospheric structure.

For the general public, however, the most spectacular results from planetary missions are the photographs, although not every spacecraft carries an imaging system. When Voyager 2 reached Jupiter in 1979 it was able to send back a maximum of 75 pictures an hour from its two cameras. Each camera system focused an image on to a selenium-sulphur light-sensitive vidicon just 11mm square. The surface was then scanned with an electron beam to break up the picture into a grid of 800 × 800 picture elements, or pixels. Each pixel was assigned a brightness level on a scale of 0–255, ranging from total black to pure white. Each of these 256 levels can be represented by an eight-digit binary number (since $256 = 2^8$). For example, 89 in binary is 01011001, which can be transmitted by a spacecraft as a series of pulses, or bits, and decoded on Earth to reconstruct a single pixel. So the total number of bits for each black-and-white picture is $8 \times 800 \times 800 = 5,120,000$. For colour, each scene has to be shot three times, through red, green and blue filters.

Operating in the neighbourhood of Jupiter, Voyager 2's transmitter was powerful enough to send back 115,200 bits a second (115.2kbits/sec), taking about 45sec for every image. Contrast this with the performance of Mariner 4 in 1965, which took a week to transmit 21 pictures at 8⅓ bits/sec!

Voyager's signal faded as it journeyed on to Saturn, where the transmission rate was limited to only 44.8kbits/sec by the ability of the radio dishes of NASA's Deep Space Network to pick out the data from background noise. At Uranus in January 1986 the rate was switched down to 21.6kbits/sec.

Giotto, the European Halley's Comet probe, had a different problem. It was not expected to survive the close encounter, so all of its colour images were transmitted in real time instead of being stored on tape for subsequent playback. Giotto was also one of the first deep-space craft to carry a charge-coupled device (CCD) camera. This solid-state silicon system has a greater spectral range and more sensitivity than a vidicon tube. The Galileo Jupiter orbiter will use one to return an image every minute at 115kbits/sec.

This is an excellent time to review the accomplishments of deep-space exploration. Galileo has been delayed by the Shuttle's troubles, as has the joint US-European Ulysses solar probe, and Soviet activity has slowed in preparation for an exciting series of missions beginning towards the end of this decade. At the same time, the Russians are displaying a new and refreshing frankness about their intentions. Europe and Japan have demonstrated their not inconsiderable prowess and can be expected to be significant participants in the major programmes of the future.

Solar system chronology

The following table presents a chronological listing of known and, in some cases, surmised planetary missions. Speculative material on early Soviet failures abounds; the flights included here are the more probable ones. There might be additions or deletions as further information comes to light. In general, the comment "tentative identification, booster failure?" indicates that although no spacecraft reached Earth orbit there is evidence that a launch took place. Beginning in 1964, Kosmos numbers were assigned to Soviet payloads stranded in Earth parking orbit. US Apollo manned lunar and Helios solar missions have been included for completeness but are not covered in the main text.

	Launch date	Nation	Launcher	Type of mission	Comments
1958					
Luna-1958A	May 1	USSR	A-1	Lunar impact?	Tentative identification; booster failure?
Luna-1958B	Jun 25	USSR	A-1	Lunar impact?	Tentative identification; booster failure?
Pioneer 0	Aug 17	US	Thor-Able	Lunar orbiter	First-stage explosion
Luna-1958C	Sep 22	USSR	A-1	Lunar impact?	Tentative identification; booster failure?
Pioneer 1	Oct 11	US	Thor-Able	Lunar orbiter	Failed to reach Earth escape velocity
Pioneer 2	Nov 8	US	Thor-Able	Lunar orbiter	Third-stage failure
Luna-1958D	Nov 15	USSR	A-1	Lunar impact?	Tentative identification; booster failure?
Pioneer 3	Dec 6	US	Juno II	Lunar flyby	Failed to reach Earth escape velocity
1959					
Luna 1	Jan 2	USSR	A-1	Lunar impact?	First lunar flyby
Pioneer 4	Mar 3	US	Juno II	Lunar flyby	Partial success; distant flyby
Luna-1959A	Jun 16	USSR	A-1	Lunar impact?	Tentative identification; booster failure?
Luna 2	Sep 12	USSR	A-1	Lunar impact	First lunar impact
Luna 3	Oct 4	USSR	A-1	Lunar flyby	First far side images
Atlas-Able 4	Nov 26	US	Atlas-Able	Lunar orbiter	Booster failure
1960					
Pioneer 5	Mar 11	US	Thor-Able	Solar probe	Successful; 0.81 × 0.99 AU solar orbit
Luna-1960A	Apr 12	USSR	A-1	Lunar flyby?	Tentative identification; booster failure?
Luna-1960B	Apr 18	USSR	A-1	Lunar flyby?	Tentative identification; booster failure?
Atlas-Able 5A	Sep 25	US	Atlas-Able	Lunar orbiter	Launcher failure
Mars-1960A	Oct 10	USSR	A-2-e	Mars flyby	Failed to reach Earth orbit
Mars-1960B	Oct 14	USSR	A-2-e	Mars flyby	Failed to reach Earth orbit
Atlas-Able 5B	Dec 15	US	Atlas-Able	Lunar orbiter	Launcher failure
1961					
Venera-1961A	Feb 4	USSR	A-2-e	Venus flyby	Failed to leave Earth orbit
Venera 1	Feb 12	USSR	A-2-e	Venus flyby	Contact lost en route
Ranger 1	Aug 23	US	Atlas-Agena B	Lunar test	High-altitude lunar test; failed to leave low orbit
Ranger 2	Nov 18	US	Atlas-Agena B	Lunar test	As Ranger 1
1962					
Ranger 3	Jan 26	US	Atlas-Agena B	Lunar lander	Launch error, lunar flyby
Ranger 4	Apr 23	US	Atlas-Agena B	Lunar lander	Spacecraft failure; first US lunar impact
Mariner 1	Jul 22	US	Atlas-Agena B	Venus flyby	Launcher failure
Venera-1962A	Aug 25	USSR	A-2-e	Venus flyby	Failed to leave Earth orbit
Mariner 2	Aug 27	US	Atlas-Agena B	Venus flyby	First Venus flyby
Venera-1962B	Sep 1	USSR	A-2-e	Venus flyby	Failed to leave Earth orbit
Venera-1962C	Sep 12	USSR	A-2-e	Venus flyby	Failed to leave Earth orbit
Ranger 5	Oct 18	US	Atlas-Agena B	Lunar lander	Spacecraft failure; close lunar flyby
Mars-1962A	Oct 24	USSR	A-2-e	Mars flyby	Failed to leave Earth orbit
Mars 1	Nov 1	USSR	A-2-e	Mars flyby	Contact lost en route
Mars-1962B	Nov 4	USSR	A-2-e	Mars flyby	Failed to leave Earth orbit
1963					
Luna-1963A	Jan 4	USSR	A-2-e	Lunar lander	Failed to leave Earth orbit
Luna-1963B	Feb 2	USSR	A-2-e	Lunar lander	Launcher failure
Luna 4	Apr 2	USSR	A-2-e	Lunar lander	Lunar miss
Kosmos 21	Nov 11	USSR	A-2-e	Venus test?	Remained in Earth orbit
1964					
Ranger 6	Jan 30	US	Atlas-Agena B	Lunar impact	Cameras failed
Luna-1964A	Feb/Mar	USSR	A-2-e	Lunar lander	Tentative identification; launcher failure?
Kosmos 27	Mar 27	USSR	A-2-e	Venus flyby	Failed to leave Earth orbit
Zond 1	Apr 2	USSR	A-2-e	Venus flyby	Contact lost en route
Luna-1964B	Apr 20	USSR	A-2-e	Lunar lander	Tentative identification; launcher failure?
Ranger 7	Jul 28	US	Atlas-Agena B	Lunar impact	Successful photographic return
Mariner 3	Nov 5	US	Atlas-Agena D	Mars flyby	Booster failure
Mariner 4	Nov 28	US	Atlas-Agena D	Mars flyby	First successful Mars flyby
Zond 2	Nov 30	USSR	A-2-e	Mars lander?	Contact lost en route

	Launch date	Nation	Launcher	Type of mission	Comments
1965					
Ranger 8	Feb 17	US	Atlas-Agena B	Lunar impact	Successful photographic return
Ranger 9	Mar 21	US	Atlas-Agena B	Lunar impact	Successful photographic return
Kosmos 60	Mar 12	USSR	A-2-e	Lunar lander	Failed to leave Earth orbit
Luna 5	May 9	USSR	A-2-e	Lunar lander	Retro failed; lunar impact
Luna 6	Jun 8	USSR	A-2-e	Lunar lander	Missed Moon
Zond 3	Jul 18	USSR	A-2-e	Lunar flyby	Photographed lunar far side
Luna 7	Oct 4	USSR	A-2-e	Lunar lander	Failed landing sequence
Venera 2	Nov 12	USSR	A-2-e	Venus flyby	Contact lost en route
Venera 3	Nov 16	USSR	A-2-e	Venus capsule	Contact lost before entry
Kosmos 96	Nov 23	USSR	A-2-e	Venus capsule?	Failed to leave Earth orbit
Luna 8	Dec 3	USSR	A-2-e	Lunar lander	Retro delay; lunar impact
Pioneer 6	Dec 16	US	Delta	Solar orbiter	Successful; 0.81 × 0.985 AU orbit
1966					
Luna 9	Jan 31	USSR	A-2-e	Lunar lander	First semi-soft landing; first surface images
Kosmos 111	Mar 1	USSR	A-2-e	Lunar orbiter?	Failed to leave Earth orbit
Luna 10	Mar 31	USSR	A-2-e	Lunar orbiter	First lunar orbiter
Luna-1966A	Apr 30	USSR	A-2-e	Lunar orbiter?	Tentative identification; booster failure?
Surveyor 1	May 30	US	Atlas-Centaur	Lunar lander	First true soft landing
Explorer 33	Jul 1	US	Delta	Lunar orbiter	Entered high Earth orbit
Lunar Orbiter 1	Aug 10	US	Atlas-Agena D	Lunar orbiter	Second lunar orbiter; photo survey
Pioneer 7	Aug 17	US	Delta	Solar orbiter	Successful; 1.012 × 1.125 AU orbit
Luna 11	Aug 24	USSR	A-2-e	Lunar orbiter	Third lunar orbiter
Surveyor 2	Sep 20	US	Atlas-Centaur	Lunar lander	Impacted
Luna 12	Oct 22	USSR	A-2-e	Lunar orbiter	Fourth lunar orbiter
Luna Orbiter 2	Nov 6	US	Atlas-Agena D	Lunar orbiter	Fifth lunar orbiter; photo survey
Luna 13	Dec 21	USSR	A-2-e	Lunar lander	Third lunar lander
1967					
Lunar Orbiter 3	Feb 5	US	Atlas-Agena D	Lunar orbiter	Sixth lunar orbiter; photo survey
Surveyor 3	Apr 17	US	Atlas-Centaur	Lunar lander	Fourth lunar lander
Lunar Orbiter 4	May 4	US	Atlas-Agena D	Lunar orbiter	Seventh lunar orbiter; photo survey
Venera 4	Jun 12	USSR	A-2-e	Venus capsule	First successful Venus atmospheric probe
Mariner 5	Jun 14	US	Atlas-Agena D	Venus flyby	Second successful Venus flyby
Kosmos 167	Jun 17	USSR	A-2-e	Venus capsule	Failed to leave Earth orbit
Surveyor 4	Jul 14	US	Atlas-Centaur	Lunar lander	Soft-landed? Contact lost
Explorer 35	Jul 19	US	Delta	Lunar orbiter	Eighth lunar orbiter
Lunar Orbiter 5	Aug 1	US	Atlas-Agena D	Lunar orbiter	Ninth lunar orbiter; photosurvey
Surveyor 5	Sep 8	US	Atlas-Centaur	Lunar lander	Fifth lunar lander
Surveyor 6	Nov 7	US	Atlas-Centaur	Lunar lander	Sixth lunar lander
Zond-1967A	Nov 21	USSR	D-1-e	Lunar flyby	Tentative identification; booster failure? See Zonds 4–8
Pioneer 8	Dec 13	US	Delta	Solar orbiter	Successful; 0.990 × 1.087 AU orbit
1968					
Surveyor 7	Jan 7	US	Atlas-Centaur	Lunar lander	Seventh lunar lander
Zond 4	Mar 2	USSR	D-1-e	Lunar flyby?	Fate uncertain; test of manned craft
Luna 14	Apr 7	USSR	A-2-e	Lunar orbiter	Tenth lunar orbiter
Zond-1968A	Apr 22	USSR	D-1-e	Lunar test?	Tentative identification; booster failure? See Zonds 4–8
Zond 5	Sep 14	USSR	D-1-e	Lunar flyby	Earth return; test of manned craft
Pioneer 9	Nov 8	US	Delta	Solar orbiter	Successful; 0.756 × 0.990 AU orbit
Zond 6	Nov 10	USSR	D-1-e	Lunar flyby	Earth return; test of manned craft
Apollo 8	Dec 21	US	Saturn V	Lunar orbiter	First manned lunar orbiter
1969					
Venera 5	Jan 5	USSR	A-2-e	Venus capsule	Successful atmospheric entry
Venera 6	Jan 10	USSR	A-2-e	Venus capsule	Successful atmospheric entry
Luna-1969A	Jan 19	USSR	D-1-e	Lunar sampler?	Tentative identification; booster failure?
Mariner 6	Feb 25	US	Atlas-Centaur	Mars flyby	Successful photographic flyby
Mariner 7	Mar 25	US	Atlas-Centaur	Mars flyby	Successful photographic flyby
Mars-1969A	Mar 27	USSR	D-1-e	Mars lander?	Tentative identification; booster failure?
Mars-1969B	Apr 14	USSR	D-1-e	Mars lander?	Tentative identification; booster failure?
Apollo 10	May 18	US	Saturn V	Lunar orbiter	Second manned lunar orbiter
Luna-1969B	Jun 4	USSR	D-1-e	Lunar sampler?	Tentative identification; booster failure?
Luna 15	Jul 13	USSR	D-1-e	Lunar sampler?	11th unmanned lunar orbiter; crashed on landing attempt
Apollo 11	Jul 16	US	Saturn V	Lunar lander	First manned lunar landing (eighth in all)
Zond 7	Aug 8	USSR	D-1-e	Lunar flyby	Earth return; test of manned craft
Kosmos 300	Sep 23	USSR	D-1-e	Lunar sampler?	Failed to leave Earth orbit
Kosmos 305	Oct 22	USSR	D-1-e	Lunar sampler?	Failed to leave Earth orbit

	Launch date	Nation	Launcher	Type of mission	Comments
Apollo 12	Nov 14	US	Saturn V	Lunar lander	Second manned landing (ninth in all)
1970					
Luna-1970A	Feb	USSR	D-1-e	Lunar sampler?	Tentative identification; booster failure?
Apollo 13	Apr 11	US	Saturn V	Lunar lander	Manned Earth return/flyby after craft failure
Venera 7	Aug 17	USSR	A-2-e	Venus capsule	Data transmitted until touchdown
Kosmos 359	Aug 22	USSR	A-2-e	Venus capsule	Failed to leave Earth orbit
Luna 16	Sep 12	USSR	D-1-e	Lunar sampler	13th unmanned lunar orbiter; first sample return
Zond 8	Oct 20	USSR	D-1-e	Lunar flyby	Earth return; test for manned craft
Luna 17	Nov 10	USSR	D-1-e	Lunar rover	14th unmanned orbiter; first rover
1971					
Apollo 14	Jan 31	US	Saturn V	Lunar lander	Third manned landing, 11th in all
Mariner 8	May 8	US	Atlas-Centaur	Mars flyby	Booster failure
Kosmos 419	May 10	USSR	D-1-e	Mars orbit/land	Failed to leave Earth orbit
Mars 2	May 19	USSR	D-1-e	Mars orbit/land	Second in Mars orbit; landing failed
Mars 3	May 28	USSR	D-1-e	Mars orbit/land	Third in Mars orbit; lander failed
Mariner 9	May 30	US	Atlas-Centaur	Mars orbiter	First in Mars orbit; photo survey
Apollo 15	Jul 26	US	Saturn V	Lunar lander	Fourth manned landing; 12th in all
Apollo 15 subsat	Aug 4	US	Apollo 15	Lunar orbiter	Released from Apollo 15
Luna 18	Sep 2	USSR	D-1-e	Lunar sampler	Failed landing attempt
Luna 19	Sep 28	USSR	D-1-e	Lunar orbiter	Returned images from orbit
1972					
Luna 20	Feb 14	USSR	D-1-e	Lunar sampler	Second successful unmanned sample return
Pioneer 10	Mar 2	US	Atlas-Centaur	Jupiter flyby	First jovian flyby
Venera 8	Mar 27	USSR	A-2-e	Venus capsule	Successful atmospheric probe
Kosmos 482	Mar 31	USSR	A-2-e	Venus capsule	Failed to leave Earth orbit
Apollo 16	Apr 16	US	Saturn V	Lunar lander	Fifth manned landing; 14th successful in all
Apollo 16 subsat	Apr 24	US	Apollo	Lunar orbiter	Released from Apollo 16
Apollo 17	Dec 7	US	Saturn V	Lunar lander	Sixth manned landing; 15th in all
1973					
Luna 21	Jan 8	USSR	D-1-e	Lunar rover	16th lunar landing; second rover
Pioneer 11	Apr 5	US	Atlas-Centaur	Jupiter/Saturn flyby	First Saturn, second Jupiter flyby
Explorer 49	Jun 10	US	Delta	Lunar orbiter	Successful
Mars 4	Jul 21	USSR	D-1-e	Mars orbiter	Retro failure; flyby
Mars 5	Jul 25	USSR	D-1-e	Mars orbiter	Fourth Mars orbiter
Mars 6	Aug 5	USSR	D-1-e	Mars lander	Contact lost before capsule landing
Mars 7	Aug 9	USSR	D-1-e	Mars lander	Missed planet
Mariner 10	Nov 4	US	Atlas- Centaur	Mercury/Venus flyby	First Mercury flyby
1974					
Luna 22	May 29	USSR	D-1-e	Lunar orbiter	Photographic survey
Luna 23	Oct 28	USSR	D-1-e	Lunar sampler	17th successful landing; sampler damaged
Helios 1	Dec 10	US/West Germany	Titan-Centaur	Solar probe	Approached to within 47 million km of Sun
1975					
Venera 9	Jun 8	USSR	D-1-e	Venus orbit/land	First Venus orbiter; first surface picture
Venera 10	Jun 14	USSR	D-1-e	Venus orbit/land	Second Venus orbiter; second surface picture
Viking 1	Aug 20	US	Titan-Centaur	Mars orbit/land	First Mars soft landing; first surface pictures; sixth orbiter
Viking 2	Sep 9	US	Titan-Centaur	Mars orbit/land	Second Mars soft landing; second surface pictures; sixth orbiter
Luna-1975A	Oct 16	USSR	D-1-e	Lunar sampler?	Failed to reach Earth orbit; booster failure
1976					
Helios 2	Jan 16	US/West Germany	Titan-Centaur	Solar probe	Approached to within 43 million km of Sun
Luna 24	Aug 9	USSR	D-1-e	Lunar sampler	18th landing; ninth sample return in all
1977					
Voyager 2	Aug 20	US	Titan-Centaur	Multiple flyby	Fourth Jupiter, third Saturn, first Uranus (+ Neptune) flybys
Voyager 1	Sep 5	US	Titan-Centaur	Jupiter/Saturn flyby	Third Jupiter, second Saturn flybys
1978					
Pioneer Venus 1	May 20	US	Atlas-Centaur	Venus orbiter	Third Venus orbiter; still operating
Pioneer Venus 2	Aug 8	US	Atlas-Centaur	Venus capsules	Five successful atmospheric probes
ISEE-3/ICE	Aug 12	US	Delta	Interplanetary observer	First cometary encounter
Venera 11	Sep 9	USSR	D-1-e	Venus lander	Successful but no surface pictures
Venera 12	Sep 14	USSR	D-1-e	Venus lander	Successful but no surface pictures

	Launch date	Nation	Launcher	Type of mission	Comments
1981					
Venera 13	Oct 30	USSR	D-1-e	Venus lander	Successful, third surface pictures
Venera 14	Nov 4	USSR	D-1-e	Venus lander	Successful, fourth surface pictures
1983					
Venera 15	Jun 2	USSR	D-1-e	Venus orbiter	Radar mapper; fourth Venus orbiter
Venera 16	Jun 7	USSR	D-1-e	Venus orbiter	Radar mapper; fifth Venus orbiter
1984					
Vega 1	Dec 15	USSR	D-1-e	Venus lander/ comet flyby	First Venus balloon; first Halley flyby
Vega 2	Dec 21	USSR	D-1-e	Venus lander/ comet flyby	Second Venus balloon; second Halley flyby
1985					
Sakigake	Jan 7	Japan	Mu-3SII	Halley flyby	Distant flyby successful
Giotto	Jul 2	Europe	Ariane 1	Halley flyby	First coma encounter
Suisei	Aug 18	Japan	Mu-3SII	Halley flyby	Flyby successful

Mating of the Viking Lander and Orbiter. Viking achieved the first Mars soft landing and surface pictures but could not provide unquestionable proof of the existence of life on the planet.

Solar system missions

Now that the total of deep space missions is about halfway through its second century, a book describing them all in equal detail would be very large indeed. Only the successful flights are described here in any depth, the failures being covered with brief notes. Chronicling the US failures is a simple matter, whereas the Soviet record is still a matter of some speculation; the descriptions of Soviet failures are based on the studies of Clark and Johnson but modification could be necessary in the coming years as further information is released. The later Zonds (4–8) are not awarded full entries since, although some returned excellent lunar photographs, they are regarded as part of the Soviet manned lunar programme. Likewise, the US Pioneers 5–8 and US/West German Helios 1–2 are not listed because they were not directed at planetary/cometary targets, the former being stationed around the Sun along the Earth's orbit to monitor solar activity and the latter venturing close in to our nearest star.

The terms *perigee* and *apogee* have been used to define the nearest and furthest points of an orbit. While probably annoying to purists, this does avoid the complication of terms such as *perilune*, *apojove*, *pericytherean* and numerous others.

Perihelion and *aphelion* have however been retained for heliocentric paths.

All of the Soviet missions have been launched from Tyuratam and those of the US from Florida. The latter are identified by launch pad number and "ETR" (Eastern Test Range). The area is now actually called the Eastern Space and Missile Center (ESMC), but since ETR was the official name during most of the period covered in this book that terminology has been adopted.

"ΔV" is used to denote change in velocity, usually in connection with course-correction or braking manoeuvres, and all times are GMT. AU stands for Astronomical Unit, the mean Earth–Sun distance of 149.598 million km. A light year is 9.4605×10^{12} km, or 63,240 AU. $1\mu m$ (a micron) $= 10^{-6}m$; 1Å (an angstrom) $= 10^{-10}m$. 1mb (a millibar) $=$ one thousandth of Earth atmospheric pressure, or $10^{-3} kg/cm^2$.

Pioneer 2 (launched November 8, 1958) is prepared atop its third stage. Note the Able stage at right.

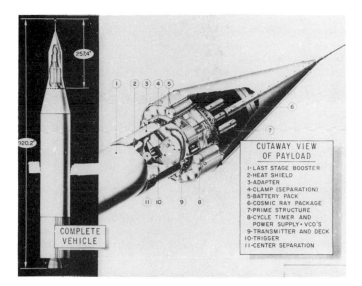

CUTAWAY VIEW
OF PAYLOAD
1-LAST STAGE BOOSTER
2-HEAT SHIELD
3-ADAPTER
4-CLAMP (SEPARATION)
5-BATTERY PACK
6-COSMIC RAY PACKAGE
7-PRIME STRUCTURE
8-CYCLE TIMER AND
POWER SUPPLY + VCO'S
9-TRANSMITTER AND DECK
10-TRIGGER
11-CENTER SEPARATION

COMPLETE VEHICLE

Above: The Pioneer 3 probe and launcher.

Right: Pioneer 4 clamped to its final stage.

Below right: Mock-up of Luna 1/2 attached to its final stage. Note the magnetometer boom and window-like micrometeoroid detector.

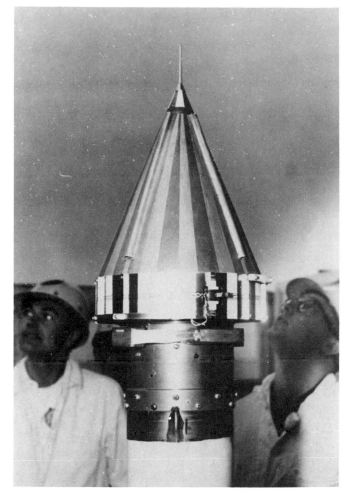

Lunas-1958A–D   The dates given for the first four Soviet lunar impact attempts in the launch chronology on page 12 are the most likely as calculated by Clark. None of the A-1 launchers penetrated deep space. Spacecraft probably identical to Lunas 1 and 2.

Pioneer 0   Mass 38kg, launched 1218 GMT August 17, 1958, by Thor-Able I on direct ascent from ETR 17A to orbit Moon with simple imaging system. Thor 127 first stage exploded after 77sec.

Pioneer 1   Mass 37.5kg, launched 0832 GMT October 11, 1958, by Thor-Able I on direct ascent from ETR 17A to orbit Moon. 250m/sec short of Earth escape speed because of error in setting second-stage cut-off. Burned up over Pacific October 13 after attaining 115,365km apogee. See Pioneer 0.

Pioneer 2   Mass 39.6kg, launched 0730 GMT November 8, 1958, by Thor-Able I on direct ascent from ETR 17A to orbit Moon. Altair third stage failed to ignite. Burned up in atmosphere after 412min flight, having reached altitude of 1,550km. See Pioneers 0 and 1.

Pioneer 3   Mass 5.87kg, launched 0545 GMT December 6, 1958, by Juno II (Jupiter No AM11) on direct ascent from ETR 5 for lunar flyby with radiation detector and camera-triggering test (there was no imaging capability). 1,030km/hr short of Earth escape speed. Attained altitude of 102,300km; burned up after 38hr 6min.

Luna 1   Launched January 2, 1959, by A-1 from Tyuratam on a direct ascent to the Moon. The 361.3kg probe, sometimes known as *Mechta* (Dream), was intended to impact the Moon (later revealed by Yuri Gagarin), but it sailed past at a distance of 5,000–6,000km and became the first man-made probe to leave the Earth's influence and enter heliocentric orbit (146.4 million × 197.2 million km), followed by the rocket's final stage. 1kg of sodium had been released from the third stage at 0057 GMT on January 3 as a tracking aid. Transmissions continued until January 5, 62hr after launch and some 600,000km from Earth. Almost identical to Luna 2.

Pioneer 4   Mass 6.1kg, launched 0545 GMT March 3, 1959, by Juno II (Jupiter No AM14) on direct ascent from ETR 5 for lunar flyby. Missed Moon by 60,000km instead of planned 32,000km. Second probe to escape Earth, first for US. See Pioneer 3.

Luna-1959A   Launched June 16, 1959, by A-1 from Tyuratam on direct-ascent trajectory for lunar impact but suffered booster failure. Tentative identification only, though US intelligence reports confirm an attempt towards mid-1959. Probably attempt at Luna 2 mission

Luna 2

Launched: September 12, 1959
Vehicle: A-1
Site: Tyuratam
Spacecraft mass: 390.2kg
Destination: Moon
Mission: Impact
Arrival: September 13, 1959 (impacted 30°N/0° longitude at 2102.14 GMT)
Payload: Magnetometer on 1m boom (sensitive to 6×10^{-4} gauss)
Micrometeoroid detector (covering 0.2m²)
Radiation detectors (including cosmic rays)
End of mission: September 13, 1959
Notes: First object on Moon

Luna 2.

Luna 2 was almost identical to its predecessor, but the A-1's launch accuracy was sufficient this time to ensure lunar impact. The Soviets did not attempt the difficult feat of establishing a lunar orbiter but were satisfied with a simple 1.2m-diameter sphere housing a set of basic instruments to investigate the almost unknown conditions beyond the Earth.

Luna 2 consisted of two polished aluminium-magnesium hemispheres bolted together at the equator and pressurised at 1.3 Earth atmospheres to help maintain an internal temperature of 20°C. While the Americans employed patterned surface finishes and thermal louvres for temperature control, the Russians supplemented these measures by housing the electronics in sealed containers. Battery-powered transmitters using the four rod antennae radioed back internal temperature and pressure values in addition to data from the instruments. There was no active attitude control, and once Luna 2 had been released from the final stage it was on its own. Since it does not even appear to have been spun up before release, the Soviet designers must have been relying on tumbling to even out solar heating of the structure.

In the 1950s the micrometeoroid population was seen as a possible hazard to manned spaceflight, and many early satellites and deep-space probes carried detectors to learn more about the numbers and distribution throughout space of these tiny high-speed particles. The piezoelectric detector on Luna 2, looking like a window on the outer surface, was divided into four areas capable of registering particles ranging in mass from 2.5×10^{-9}g to 2×10^{-7}g and above. Luna 2 and the third stage also carried 9cm and 15cm metallic spheres bearing the Soviet emblem for deposition on the lunar surface.

A batch of sodium was ejected from the third stage as a tracking aid and an indicator of deep-space conditions. It expanded to a cloud 650km in diameter before fading from sight.

As Luna 2 and its upper stage plummeted towards the Moon, the magnetometer failed to detect any field 55km above the surface and there was no sign of a radiation belt. Travelling at 3.3km/sec, the probe ploughed into the surface 30° from the vertical on the edge of Mare Imbrium near Archimedes. Soviet astronomers claimed to have seen the impacts through telescopes – hard to believe in view of the fact that the impacts of the vastly larger Saturn V third stage during the Apollo programme went unobserved – and assurances were given that both vehicles had been sterilised to avoid contaminating another world with terrestrial bacteria.

Luna 3

Launched: October 4, 1959
Vehicle: A-1
Site: Tyuratam
Spacecraft mass: 278.5kg
Destination: Moon
Mission: Flyby
Arrival at target: October 6, 1959 (closest approach ~6,000km at 1416 GMT)
Payload: Photographic-TV imaging system
Micrometeoroid detector
Cosmic ray detector
End of mission: Contact lost November 1959. Re-entered Earth's atmosphere April 20, 1960
Notes: First lunar far-side pictures

Luna 3's success in late 1959 was the high point of the early era of planetary exploration, surpassed only by Mariner 2 in 1962. It not only completed a close flyby of the Moon and returned the first images of the far side, but did so at the first attempt. This was a remarkable achievement for contemporary technology: the launch chronology illustrates the large number of lunar impact, flyby and orbiter failures by both space powers. The probe's size and the accuracy of its trajectory were in stark contrast to those of US ventures.

Luna 3, 1.3m long and with a main body diameter of 1.19m, was based on a pressurised cylindrical shell capped at either end by hemispheres. For the first time a Soviet spacecraft carried solar cells to power equipment and charge batteries, greatly improving on the lifetimes of a few days possible with the first Lunas. (It was not until Luna 17 in 1970 that a Soviet probe again utilised solar cells.) Rectangular areas between cell strips on the cylindrical body carried "thermal shutters" to help maintain internal temperatures at 25–30°C, the 230mb

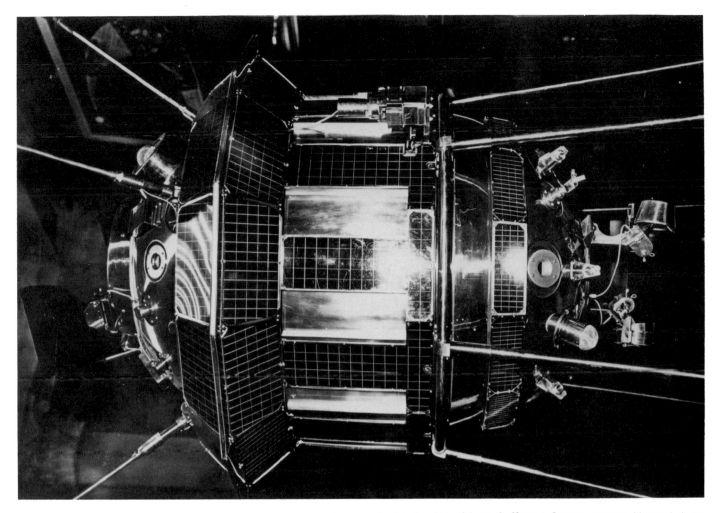

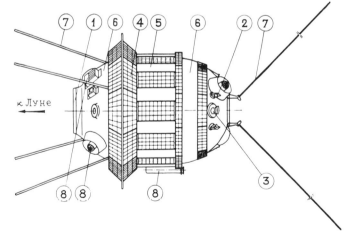

к Луне

*Soviet drawing of Luna 3. **Key: 1** Camera system (thermal doors closed) **2** gas-jet nozzles **3** solar sensor **4** solar cells **5** thermal shutters **6** thermal protection **7** antennae **8** scientific instruments.*

atmosphere being force-circulated by fans to transfer heat generated by the electronics to the walls. Thermal control was aided by spin stabilisation during Earth–Moon flight.

Since there was no provision for mid-course correction (introduced on the second-generation Lunas), active radio guidance of the launcher's third stage was used to establish a precise trajectory. Luna 3 passed over the Moon's southern polar regions and was perturbed northwards over the far side. As it climbed up above the Earth-Moon plane, ground control

ordered gas jets at one end to stop the spin. A Sun sensor on this same end locked on to the bright solar disc and the jets fired to orientate Luna 3 so that a second sensor at the other, camera, end could locate the Moon. The probe was now aligned to within 0.5–0.7° of the Moon-Sun axis, allowing the two lens systems to view the surface. A 200mm, f/5.6 wide-angle lens, for capturing the whole disc, and a 500mm, f/9.5 lens capable of higher resolution produced images 10mm and 25mm diameter respectively. The changing orientation of the images showed that Luna 3 still had a slow spin.

At 0330 GMT on October 7, 65,200km above the 3°-diameter Moon, Luna 3 began taking the first of at least 29 exposures over a 40min period using radiation-hardened 35mm photographic film. It then began spinning again and ground controllers ordered the strip of film to be developed, fixed and dried in an automatic processor, ready to be scanned by a light beam at up to 1,000 lines per image for transmission to Earth. The system was more complex than a conventional TV system and limited by the film supply, but it produced higher-resolution

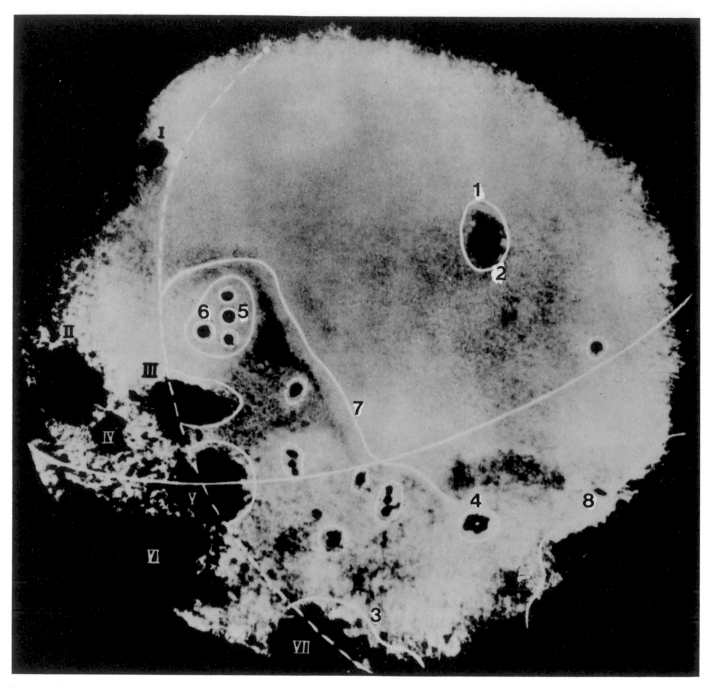

images. It was not until their Lunar Orbiters began systematic lunar mapping in the mid-1960s that the Americans used a similar system. The method was also adopted for use aboard Earth-orbiting military reconnaissance satellites, Luna 3 having clearly demonstrated the technology.

The lunar encounter did not throw Luna into a heliocentric path. It remained instead in an elliptical Earth orbit stretching out to the Moon, reaching its first apogee of 480,000km on October 10 and returning to within 47,500km of the Earth eight days later. It is not clear whether the negatives were scanned and transmitted to Earth soon after Luna 3 cleared the Moon or on October 18 near perigee. The confusion perhaps arose when the Soviets noted that Luna 3 had two data-transmission rates: low for cislunar operation and high for near-Earth. This

could also account for the story of a failed second scanning and transmission session closer to Earth in an effort to achieve greater signal strength.

Whenever the images were received, the Soviets acquired pictures of 70% of the previously unseen lunar far side with only their sixth satellite. (Zond 3 imaged the remaining 30% in 1965.) Because of the requirements of the orientation system, the Sun had been directly overhead at the time. Surface contrast was therefore unavoidably low, though mare and some crater-like features were visible. Mare Moscovrae (Moscow Sea) and Mare Desiderii (Sea of Dreams) were named, as was the prominent lava-filled Tsiolkovsky crater. Despite their low quality (prompting some US politicians and writers to claim they were fakes), the pictures did show there were far fewer

*Left: A montage of Luna 3 images revealed the lunar far side for the first time. The solid line is the equator, the broken line the limit of visibility from Earth. **Key: 1** Sea of Moscow **2** Gulf of Cosmonauts **3** Mare Australe **4** Tsiolkovsky crater **5** Lomonosov crater **6** Joliot-Curie crater **7** Sovietsky mountain range **8** Sea of Dreams. The Roman numerals indicate near-side features: **I** Mare Humboldt **II** Mare Crisium **III** Mare Marginis **IV** Mare Undarum **V** Mare Smythii **VI** Mare Foecunditatis **VII** Mare Australe.*

mare areas on the far side, requiring revision of existing theories of lunar evolution.

Atlas-Able 4   Mass 169kg, launched 0726 GMT November 26, 1959, by Atlas-Able (Atlas No 20D) on direct ascent from ETR 14 as lunar orbiter with TV scanner. Payload shroud failed after 45sec. When the project was initiated in November 1958 the plan had been to launch two Venus orbiters during the June 1959 window, but the attempts were cancelled following early Soviet lunar successes. See Pioneer 5.

Pioneer 5   Mass 43kg, launched March 11, 1960, by Thor-Able from ETR 17A towards near-Venus space to demonstrate deep-space probe technology. Originally intended for Venus flyby, launch June 1959, but mission downgraded to solar orbit. Final telemetry received June 26, 1960, at record Earth distance of 36.2 million km. Orbit 0.806 × 0.995 AU, 312 days.

Luna-1960A and -1960B   US intelligence reports indicate that two Lunas launched by A-1 from Tyuratam failed to reach Earth orbit in the second quarter of 1960. Launches on April 12 and 18 would have permitted flyby/impact missions; the probes possibly carried improved Luna 3-type camera systems. The next Soviet lunar mission was not launched until 1963, by which time the more powerful A-2-e vehicle was in service.

Atlas-Able 5A   Mass 175.5kg, launched 1513 GMT September 25, 1960, by Atlas-Able (Atlas No 80D) on direct ascent from ETR 12 as lunar orbiter. Second stage shut down prematurely and payload burned up after 17min. See Atlas-Able 4.

Above. Launch of Atlas-Able 5A on September 25, 1960. All three of the series failed.

Mars-1960A   Launched October 10, 1960, probably by A-2-e from Tyuratam. First use of larger booster and Earth parking orbit. Mars flyby in May 1961 by 850?kg craft (though possibly Luna 1-type impact craft launched by A-1) intended, but third stage failed and payload did not attain Earth orbit. Failure announced by US in September 1962.

Mars-1960B   Launched October 14, 1960, by A-2-e from Tyuratam. Other comments as Mars-1960A.

Atlas-Able 5B   Mass 176kg, launched 0840 GMT December 14, 1960, by Atlas-Able (Atlas No 91D) on direct ascent from ETR 12 as lunar orbiter. Launcher exploded after 68sec. See Atlas Able 4 and 5A.

Venera-1961A   Launched February 4, 1961, by A-2-e from Tyuratam into 224 × 328km, 65°, 89.8min Earth parking orbit (first for a planetary craft). Re-entered February 26. Believed to have been intended as first Venus flyby probe. Sometimes known as Sputnik 7. See Venera 1.

Venera 1   Mass 643.5kg, launched February 12, 1961, by A-2-e from Tyuratam for Type I Venus flyby. Contact lost on February 27 with craft some 23 million km from Earth following departure from parking orbit. Inactive Venus flyby at a distance of ~100,000km on May 19/20, 1961.

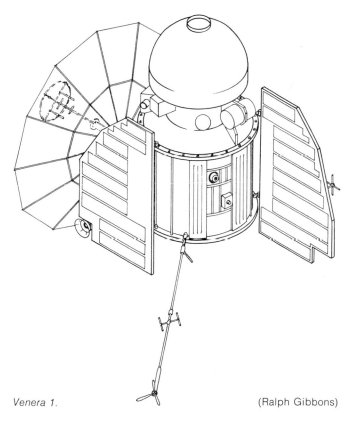

Venera 1.

(Ralph Gibbons)

RANGER 3 SPACECRAFT

Ranger 3, with the hard-landing capsule mounted atop a solid-propellant retromotor. None of these early landers (Rangers 3–5) succeeded. A later type of capsule was planned for Rangers 10–15 to carry a surface TV camera, but these "Block V" Rangers were cancelled in December 1963.

Ranger 3 Mass 330kg, launched 2030 GMT January 26, 1962, by Atlas-Agena B from ETR 12 for impact mission with TV camera and 44kg hard-landing capsule with seismometer. Launcher fault produced a Moon miss distance of 32,000km and incorrect course correction led to a 36,785km flyby at 2323 on January 28. The Central Computer and Sequencer failed and no TV images were returned. See Ranger 7.

Ranger 4 Mass 331kg, launched 0850 GMT April 23, 1962, by Atlas-Agena B from ETR 12 with TV and landing capsule. Central Computer and Sequencer failure left it drifting aimlessly. Crashed at 15.5°S/130.5°W on lunar far side at 0949 on April 26. See Ranger 7.

Mariner 1 Mass 204kg, launched 0921 GMT by Atlas-Agena B from ETR 12. Intended to fly past Venus at 29,000km on December 8, 1962, but incorrect Atlas 145D trajectory forced range safety destruct after 293sec. See Mariner 2.

Venera-1962A Launched August 25, 1962, by A-2-e from Tyuratam for Venus flyby/impact (?) but failed to leave Earth parking orbit. Reported by official US sources.

Mariner 2

Launched: 0653 GMT August 27, 1962
Vehicle: Atlas-Agena B (Atlas No 179D)
Site: ETR 12
Spacecraft mass: 203.6kg at launch
Destination: Venus
Mission: Flyby
Arrival at target: December 14, 1962 (closest approach 34,827km)
Payload: Microwave radiometer
Infra-red radiometer
Magnetometer
Ion chamber
Cosmic dust detector
Solar plasma analyser
Particle flux detector
End of mission: Last contact January 3, 1963
Notes: First successful Venus flyby

The early US planetary (ie non-lunar) exploration schedule depended on the new cryogenic Centaur stage becoming available for use with the Atlas booster, producing a capability of 567kg to Venus or Mars. Development delays pushed Centaur's planetary debut back to 1969 with Mariners 6 and 7. JPL officials decided that a Venus flyby probe based on the lunar Ranger craft and launched by Atlas-Agena B could carry at least 9kg of scientific instruments during the 56-day July 18–September 12, 1962, launch window. This opportunity would otherwise have been lost, forcing a wait until 1964 at least.

The plan was approved by NASA in early September 1961 – only a year before launch – and Mariner R's (R for Ranger) objectives were simply set: "To pass near the planet, to communicate with the spacecraft from the planet, and to perform a meaningful planetary experiment." This at a time when the Ranger series was failing in its attempts at the Moon.

The seven instruments on each of two craft would study particle radiation and magnetic fields at the planet and during the four-month interplanetary cruise period. A scan mechanism would carry a microwave radiometer (working at wavelengths of 19 and 13.5mm) to measure Venus' temperature close to the surface and an infra-red (IR) radiometer to look at the thermal characteristics of the higher atmosphere in the 8–9 and 10–10.8μm wavebands. There was no imaging capability; the first US Venus probe to be so equipped would be Mariner 10 in February 1974.

Like Ranger, Mariner consisted of a basic magnesium and aluminium framework topped by an aluminium superstructure. The base hexagon provided side mounting points for the six electronic bays (control, telemetry, etc) and a housing for the central course-correction motor. The complete 222N thruster weighed 16.8kg when fully fuelled with anydrous hydrazine and could be fired once for 0.2–57sec to provide a total ΔV capability of up to 61m/sec.

All of the experiments except the solar plasma detector were mounted on the superstructure, with the magnetometer nestling just below the omni antenna at the top. This antenna, one of four carried by each Mariner R, was used to transmit telemetry until the course correction had been carried out. An encounter data rate of 8⅓ bits/sec at 3W was possible with the 48.9cm-diameter Earth-pointing high-gain antenna mounted on the base. Two control antennae were carried on one solar panel to receive commands for manoeuvres and other activities. Provision for attitude stabilisation during the cruise comprised Earth and Sun sensors and 10 jets drawing on 1.95kg of nitrogen (sufficient for 200 days); the sensors kept the superstructure pointing directly Sunwards as the spacecraft rolled around its long axis.

Power was provided by two rectangular solar panels feeding a 15kg silver-zinc 1kW-hr battery. The panels were originally 1.52 × 0.76m rectangles – each with 4,900 cells and together capable of producing a total of 148–222W – but one was modified to add 910 cells at one end. A counterbalancing "solar sail" was added to the other panel to help with attitude control using the pressure of sunlight (Mariner 4 had a similar system).

Approaching Venus, the Sun's heat load would almost double, calling for careful protection of the electronics. External surfaces carried paint patterns, gold plating, polished aluminium and thermal louvres to ward off the heat. The rear faces of the solar panels were painted black to radiate away heat, and the base hexagon carried shields on its upper and lower surfaces. These thermal control measures accounted for 4.5kg of Mariner's total 203.6kg.

The 54,000 components were packed into a spacecraft that stood 3.02m high at launch and 3.63m during cruise with the main antenna deployed, and 5.03m across the panels. Contact would have to be maintained for 2,500hr, even though JPL's experience with Ranger stretched to only 65hr per mission.

Mariner R-2 was brought out of storage and launched 36 days after the launch failure of its twin (see Mariner 1). The Atlas booster again almost caused the mission to be aborted when one of its side-mounted vernier steering thrusters went over hard to one side for 60sec and generated a launcher roll

The solar panels are attached to Mariner 2. At the top of the tower is the omni antenna, with the small cylindrical magnetometer just below. Halfway down the tower, the black spherical ion chamber rests adjacent to the Geiger tubes. Opposite is the large radiometer dish, with its two reference horns above. The device resting on the base directly below the ion chamber is the cosmic dust detector.

rate of 1rpm. Control had been regained by the time of upper-stage separation but the Agena's horizon scanner was taken to within 3° of its operating limits.

Mariner 2/Agena coasted in a 187km parking orbit for 16.3min before a second Agena burn injected the spacecraft into a heliocentric trajectory at 40,900km/hr 26min 3sec after launch. Eighteen minutes later the solar panels extended from their upright position to produce 195W of power. On October 31 the output of one panel fell abruptly – probably as a result of a short circuit – but the other alone was sufficient for the Venus encounter.

Following the second Agena burn Mariner 1 looked set to miss the planet by 375,900km, compared with an acceptable 12,900–64,000km and a desirable 29,000km approaching from the night side. A course correction of 112km/hr was required. On September 4 the hydrazine motor fired through the base of the hexagonal mainframe for 28.3s, resulting in a calculated Venus miss distance of 40,990km after a journey of 109.5 days. Impact had to be avoided since the craft was unsterilised.

By September 12 Mariner 2 was 4,310,447km from Earth, and on November 6 it became equidistant between Earth and Venus (22,365,000km). The cruise experiments returned data during this phase, the two radiometers being switched on only for the encounter. The craft approached Venus from the night side on December 14 and the radiometers' scan platform began to nod the 48.9cm-diameter dish at 0.1°/sec through 120° to execute three scans: the first on the night side, the second across the terminator and the last on the day side.

The radiometers indicated a surface temperature of at least 425°C, with little difference between the day and night hemispheres. In addition, a dense cloud layer extended from 56km to 80km above the surface. The magnetometer and the radiation detectors produced no signs of a magnetic field or of any radiation trapped in one. Nor did the solar plasma probe detect any change in the incoming solar wind. Combined with the relatively large flyby distance, these factors suggested a magnetic field 10% the size of Earth's at most. Closer flybys would be necessary to refine the figure. The 0.84kg dust detector, capable of sensing impacts by 10^{-9}g objects, registered no events during the 35min encounter and only two throughout the mission.

Mariner 2's path was deflected 40° by Venus and the probe continued to return interplanetary cruise data. The plasma instrument, for example, showed the solar wind to be a steady outstreaming of material from the Sun. Last contact was made at 0700 GMT on January 3, 1963, at the record Earth distance of 86.7 million km. The craft's 129 days of continuous operation also represented a new record, comfortably exceeding what had seemed attainable when Mariner 2 had been launched on a Type I trajectory to keep the flight as short as possible.

Mariner 2 was so successful that a follow-up mission by the R-3 craft (assembled from spares) in the 1964 window was cancelled in January 1963.

Venera-1962B and -1962C Launched September 1 and 12, 1962, by A-2-e from Tyuratam for Venus flyby/impact (?) but failed to reach Earth parking orbit. Report by official US sources.

Ranger 5 Mass 342.4kg, launched 1759 GMT October 18, 1962, by Atlas-Agena B from ETR 12 with TV and landing capsule. All solar panel power was lost and the battery was exhausted before the course-correction burn. Missed Moon by 724km on October 21. See Ranger 7.

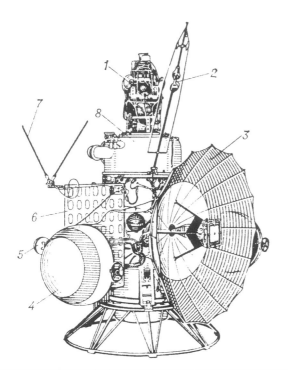

Mars-1962A Mass 894kg? Launched October 24, 1962, by A-2-e from Tyuratam but failed to leave Earth parking orbit. See Mars 1.

Mars 1 Mass 893.5kg, launched November 1, 1962, by A-2-e from Tyuratam. Intended to photograph Mars as it flew by at up to 11,000km. Contact was lost March 21, 1963, possibly because of attitude-control problems. By then Mars 1 had broken Mariner 2's communications distance record. Course correction was not made and it flew past Mars at 193,000km around June 19. Craft similar to Venera/Zond late first-generation vehicles.

Mars-1962B Mass 894kg? Launched November 4, 1962, by A-2-e from Tyuratam but failed to leave Earth parking orbit. See Mars 1.

Luna-1963A Launched January 4, 1963, by A-2-e from Tyuratam but failed to leave Earth orbit. Revealed by US intelligence in September 1962, it was the first of the second-generation Lunas, probably of the Luna 9 lander type.

Left: Mars 1. **Key:** *1 propulsion system 2 magnetometer 3 high-gain antenna 4 hemispherical thermal-control radiator 5 antenna 6 solar panel 7 omni antenna 8 pressurised compartment. Framework at bottom is ground handling equipment.*

Below: *Ranger 6 is prepared for launch.*

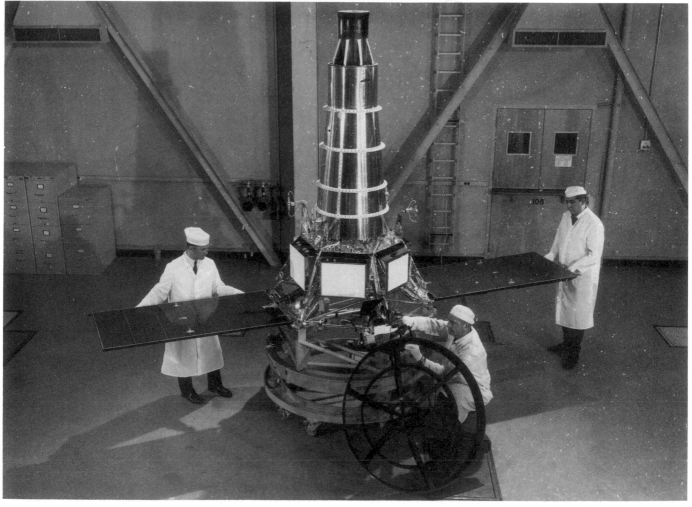

Luna-1963B Launched February 2, 1963, by A-2-e from Tyuratam but failed to reach Earth orbit. US intelligence reports indicate a total failure in February; the date given here satisfies a Luna 9-type lander mission.

Luna 4 Launched April 2, 1963, by A-2-e from Tyuratam into 167 × 297km, 64.7° parking orbit before injection into trans-lunar path. Mass given as 1,422kg. Probably landing attempt of Luna 9 type but course correction appears to have failed, resulting in flyby at 8,500km on April 6. Entered barycentric (ie Earth-Moon) orbit and finally perturbed into heliocentric path.

Ranger 6 Mass 364.6kg, launched 1549 GMT January 30, 1964, by Atlas-Agena B from ETR 12 after lengthy investigations into previous failures. First Ranger to carry block of six TV cameras, but no pictures were returned before impact at 9.39°N/21.51°E at 092432 on February 2; TV power supply had been shorted out during Atlas booster separation. See Ranger 7.

Luna-1964A and -1964B Launched February/March and April 20, 1964, respectively by A-2-e from Tyuratam but failed to reach Earth orbit. Two launch failures were indicated by US intelligence reports. The second date was calculated by Clark on the assumption that they were Luna 9-type lander missions.

Kosmos 27 Launched March 27, 1964, by A-2-e from Tyuratam for Venus flyby but failed to leave Earth parking orbit and burned up the following day. See Zond 1.

Zond 1 Mass ~825kg, launched April 2, 1964, by A-2-e from Tyuratam for Type I Venus flyby but contact was lost mid-May following two course corrections. Venus flyby of ~100,000km on about July 19 was intended, with an impact possible.

Ranger 7

Launched: 1650 GMT July 28, 1964
Vehicle: Atlas-Agena B (Atlas No 250D, Agena 6009)
Site: ETR 12
Spacecraft mass: 366kg
Destination: Moon
Mission: Impact
Arrival at target: July 31, 1964
Payload: Package of six TV cameras
End of mission: July 31, 1964, impact at 1325 GMT
Notes: First Ranger success, first high-resolution lunar images

Ranger 6 was the first of the Block III Ranger craft, which had a single objective: to obtain high-resolution pictures of the Moon in support of Project Apollo. Miscalculation of the payload capacity of early Atlas-Agena Bs had led JPL to build Rangers 3–5 without back-up systems in order to keep the weight down.

The original basic spacecraft consisted of a magnesium 152cm-diameter hexagonal framework containing the electronics (including the Central Computer and Sequencer) and providing attachment points for two solar panels and a 1.22m-diameter high-gain antenna. The 1.85m-long trapezoidal panels each carried 4,340 cells to generate a total of 155–210W, with an 11.3kg silver-zinc battery providing a 1kW-hr capacity. Three-axis attitude control was maintained by 10 nitrogen jets mounted on the main body and supplied from three tanks holding 1.1kg of the pressurised gas, with six Sun and two Earth sensors and three gyros providing attitude

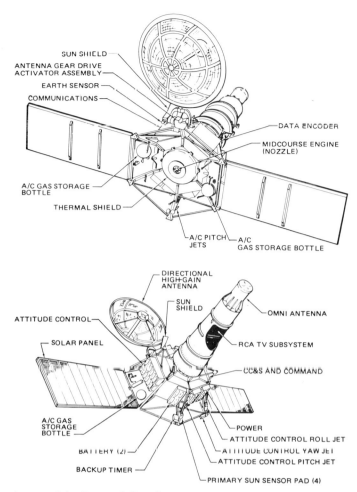

Layout of the Ranger 6–9 craft.

reference. In cruise mode Ranger's long axis pointed directly at the Sun, keeping the panels square-on to the solar radiation.

A 16.3kg mid-course correction unit was housed in the centre of the main body, with the 220N hydrazine motor, capable of burning for a maximum 68sec to produce a ΔV of 44m/sec, pointing away from the hexagonal base.

Ranger carried a burden that was to prove troublesome, and sometimes fatal, throughout the programme: sterilisation. The lunar probe was deliberately overdesigned for its purpose (for example, it could have used batteries instead of solar panels) so that it could act as a testbed for planetary missions. NASA ordered all planetary/lunar vehicles to be sterilised to avoid contaminating other worlds, although there were few worries about the harsh lunar environment. Nevertheless, all of Ranger's components were heated to 125°C for 24hr, its surface cleaned with alcohol, and the whole craft bathed in ethylene oxide gas as it nestled under the Agena shroud on the launch pad. Investigations later showed that sterilisation damaged components and contributed to the string of failures, particularly Ranger 3.

Beginning with Ranger 6, the new vehicles included a second battery, a larger course-correction capacity (60m/sec), a second nitrogen attitude-control system, Mariner 2-type rectangular solar panels with 9,792 silicon cells, and

Above: *The six-camera RCA package seen before insertion into Ranger 7.*

Below right: *Ranger 7 provided the first close-up views of the lunar surface. The area covered here is about 9km². An Apollo 16 picture, taken in 1972, revealed that the actual impact crater was outside the circle predicted from Ranger's own images.*

aluminium instead of magnesium for the framework (heavier but with better thermal characteristics). The Central Computer and Sequencer was built of components that had not been heat-sterilised and carried additional features (such as timers) to allow a basic mission in the event of equipment failure.

The RCA 173kg TV unit was housed above the main body where the landing capsule and its retromotor had been sited on the Block II Rangers 3–5. The 1.5m-high tower carried an outer layer of polished aluminium for thermal protection, and tapered to 41cm diameter from the 68.6cm at the base. A 33cm-square cutout in the side allowed six cameras to view the lunar surface during descent. Two full-scan cameras – F1 (25mm, f/1) and F2 (75mm, f/2) – recorded 1/200th-sec images every 2.56sec on 2.8cm² vidicons. These were scanned into 1,152 lines for transmission to Earth. Four partial-scan cameras reduced 1/50th-sec images on 0.7cm² vidicons into 300 lines: P1 (75mm, f/2), P2 (75mm, f/2), P3 (25mm, f/1) and P4 (25mm, f/1).

Ranger 7 became the second major success of the US planetary/lunar programme – Mariner 2 was the first – following 13 consecutive lunar failures. It also marked a turning point; to this day there have been few more failures in the entire US deep-space programme.

The impact point selected for Ranger 7 was at 10.7°S/20.7°W, on the northern rim of the Sea of Clouds in a region crossed by rays from Copernicus. It had been suggested that Ranger should go instead to the Ranger 6 site, where that craft's fresh impact crater would provide data on the nature of the lunar surface. Unfortunately, the lighting conditions were now inappropriate (the target had to be near the terminator for good surface feature contrast).

The first Agena burn established a 185km parking orbit and the second produced a final velocity of 39,461km/hr, within 6½km/hr of the planned figure. The launch had been so accurate that even without a course correction Ranger would have skimmed past the Moon's leading edge and struck the far side. At 102709 GMT on July 29 a 50sec correction burn shifted the impact point to within 16km of the planned point.

Twenty minutes before impact, ground signals commanded the two full-scan cameras to begin their 90sec warm-up, followed by the four remaining cameras five minutes later. JPL's Goldstone dish in California received strong video signals all the way down as Ranger raced to a 9,413km/hr collision, ground equipment recording the 4,316 pictures on videotape and 35mm film.

Project scientists were delighted with the results: the resolution of the last images was 1,000 times better than any obtainable from Earth. The surface had clearly been comprehensively pounded in its history and there were a large number of small craters. Nevertheless, Apollo officials declared that a manned landing would be easier than previously believed, the mare region being generally smooth with gentle slopes. Ridges and flow patterns exhibited typical lava features. The Apollo 16 orbital cameras found the 14m-diameter impact craters of Rangers 7 and 9 in 1972.

Mariner 3 Mass 261kg, launched 1922 GMT November 5, 1964, by Atlas-Agena D from ETR 13 for first US Mars flyby. Shroud failed to separate and spacecraft sailed off into heliocentric orbit, battery power failing 8hr 43min after launch.

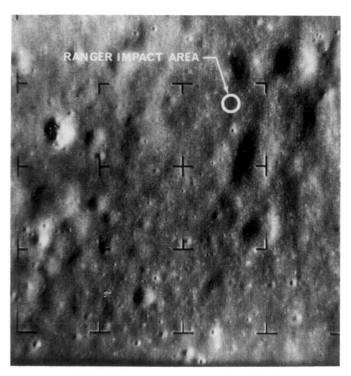

Mariner 4

Launched: 1422 GMT November 28, 1964
Vehicle: Atlas-Agena D (Atlas No 288D)
Site: ETR 12
Spacecraft mass: 261kg at launch
Destination: Mars
Mission: Flyby
Arrival: July 15, 1965, closest approach of 9,600km at 0101 GMT
Payload: TV system (5.1kg)
Meteoroid detector (0.95kg)
Cosmic ray telescope (1.2kg)
Ionisation chamber (1.3kg)
Magnetometer (3.1kg)
Trapped radiation detector (1.0kg)
Solar plasma probe (2.9kg)
End of mission: Formally October 1, 1965, last contact December 21, 1967
Notes: First successful Mars flyby and return of images

The two probes of the Mariner-Mars 1964 project, approved by NASA in November 1962, began the US assault on the Red Planet. The large Atlas-Centaur was not yet available but the smaller Atlas-Agena D was powerful enough for a basic mission: a single TV camera on a scan platform with an f/8, 30.5cm-focal-length Cassegrain telescope would return up to 21 pictures of the planet's surface as it flew past following an eight-month journey from Earth. Each frame would take 24sec to record on tape from the slow-scan vidicon tube, and fully 8hr 20min to transmit to Earth at 8⅓ bits/sec via the 117 × 53cm elliptical 10.5W S-band antenna.

The main body consisted of a 138.4cm-diameter, 45.7cm-high octagonal magnesium frame containing seven compartments of electronics and an eighth with a 220N hydrazine course-correction system rated at a maximum ΔV of 81m/sec. The design allowed only two firings at most. Since Mariner-Mars would travel away from the Sun, it carried four solar panels producing 700W of power at Earth. Each had 7,056 cells covering a total of 6½m², and the whole array fed a 1,200W-hr silver-zinc battery. Three-axis attitude control was achieved by 12 nitrogen jets mounted on the solar panel tips, a gyro system and Sun and Canopus sensors (the first use of a star sensor aboard a US deep-space probe). Four solar vanes totalling 0.65m² were carried on the end of each panel to help provide attitude control using solar pressure. Total spacecraft span was 6.79m.

Analysis of the Mariner 3 failure indicated that the inner fibreglass layer had separated from the payload shroud's outer skin and prevented ejection. The second mission was postponed while engineers worked around the clock for 17 days to build and qualify an all-metal shroud. The Agena D's first burn took Mariner into a 172 × 184km, 28.3° parking orbit. Then, following a 1,925sec coast, a 95sec burn resulted in escape at 11.50km/sec into a Type I Mars trajectory.

The spacecraft separated from its upper stage 2,708sec after launch, heading for a Mars miss distance of 246,378km on July 17, 1965. The single course-correction burn was executed at 1609 GMT on December 5, when Mariner 4 was 2,034,000km from Earth, the 220N thruster firing for 20sec to yield a ΔV of 17.3m/sec against the planned 17.0m/sec. This switched the flyby to 9,600km above the surface on the other side of the planet on July 15.

Most of the experiments were active during the cruise phase. Though this was a period of low solar activity no fewer than 20 solar flares were detected, and the dust instrument picked up about 235 impacts through to the encounter. The ionisation chamber failed in February and only half of the data from the solar plasma probe were usable

Launch of Mariner 4 by Atlas-Agena D.

Mariner Mars 1964

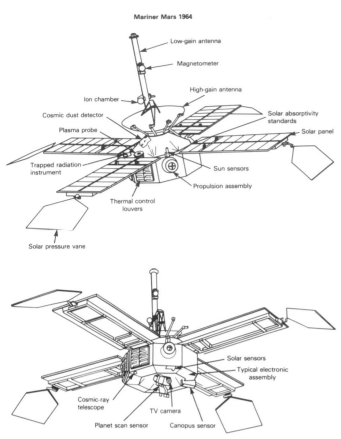

Above: The Mariner 4 spacecraft.

Below: Picture 11 returned by Mariner 4 clearly showed craters at 33°S/197°E in the Atlantis region. This was taken at 003033 on July 15, 1965, from a slant distance of 12550km. North is at top.

Forty minutes before closest approach, the TV system began an automatic 25min picture-taking sequence, starting at about 37°N on the left-hand limb of the visible disc near Phlegra and sweeping down through the Mare Sirenum to about 55°S. A total of 21 partially overlapping images, taken through red and green filters, plus 22 lines of a 22nd were recorded on 100m of tape for subsequent playback. Some 1¼hr after closest approach, Mariner 4 dipped behind the right-hand limb as viewed from Earth so that its radio signals were refracted through the atmosphere. The fluctuations indicated a surface pressure of about 5mb, whereas anything up to 80mb had been expected. This meant that Mars landers would have to carry retrorockets in addition to parachutes. Surface temperatures were put at around −100°C during the day. The magnetometer placed an upper limit on the magnetic field of 0.1% of Earth's, a finding supported by the lack of radiation belts analogous to the Earth's van Allen regions.

Yet more surprises were in store when the surface images began to be played back the next day. The 40,000 elements of each picture meant that the replay lasted from July 15 to 24. The first image, recorded from a distance of 16,900km, showed atmospheric haze on the limb. Roundish patches became evident in subsequent pictures until Frame 7, produced at a slant range of 13,500km, revealed definite craters. This was a surprise and, in conjunction with the thin atmosphere, showed Mars to be an ancient Moon-like body unlikely to harbour life. About 70 craters with diameters of 5–120km and rims sometimes reaching up to 100m above the surface were registered in Frames 5–15. Frames 19–22 crossed over the terminator on to the night side and showed no details.

The post-Mariner 4 picture of Mars was a depressing one for exobiologists. But the small probe had imaged only one per cent of the surface and it was not until Mariner 9 in 1972 that the planet was seen to possess a rich diversity of features, some of them probably created by running water.

The last telemetry of the encounter was received at 2205 GMT on October 1, 1965, as Mariner 4 moved beyond 309 million km from Earth. It circled the Sun and contact was re-established in 1967. On September 15, 1967 it apparently passed through a meteoroid shower 47.6 million km from Earth, detecting 17 strikes in 15min. Attitude was perturbed slightly and a temperature drop of 1°C indicated damage to the thermal protection.

In October 1967 the craft was used for attitude-control tests in support of Mariner 5 (the Mariner 4 back-up modified for a Venus mission), and at 060306 GMT on October 26, after 1060 days in space, the main engine was fired for 70 sec, producing a ΔV of 62m/sec. This record-breaking reliability was further demonstrated when portions of Frames 16 and 17 were played back without any sign of degradation. The TV system also successfully imaged black space on November 22. On December 7 the supply of control gas was finally exhausted and the solar panels could no longer maintain Sun-lock to charge the battery. Ten days before the final communications session – on December 21, 1967 – 83 further meteoroid hits were recorded.

Zond 2 Mass 1,145kg (?), launched November 30, 1964, by A-2-e from Tyuratam. Contact lost April 1965. Mars photographic flyby probably intended; a 1,500km encounter was achieved on August 6, 1965. A landing capsule might have been included: Zond 3 (see entry) was probably intended originally for this Mars window and is known to have been very similar to Venera 3. It also seems to have been launched to minimise arrival speed, so reducing the problems of atmospheric entry.

Ranger 8

Launched: 1105 GMT February 17, 1965
Vehicle: Atlas-Agena B (Atlas No 196D)
Site: ETR 12
Spacecraft mass: 366.4kg at launch
Destination: Moon
Mission: Impact
Arrival at target: February 20, 1965
Payload: See Ranger 7
End of mission: February 20, 1965, impact at 0957 GMT
Notes: Second successful US lunar imaging mission

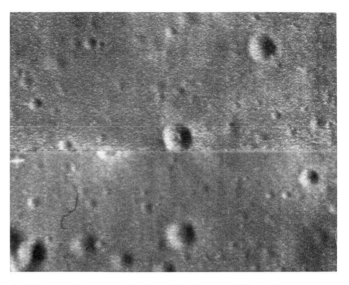

Left: Ranger 8's camera B returned this image of 50km-diameter crater Delambre (centre) from a height of 760km about 7min before impact.

Above: The impact crater formed by Ranger 8 was imaged by Lunar Orbiter 2. Note the central peak in the 13 × 14m crater.

Ranger 9

Launched: 2137 GMT March 21, 1965
Vehicle: Atlas-Agena B (Atlas No 204D, Agena 6007)
Site: ETR 12
Spacecraft mass: 366.8kg at launch
Destination: Moon
Mission: Impact
Arrival at target: March 24, 1965
Payload: See Ranger 7
End of mission: March 24, 1965, impact at 1408 GMT
Notes: Third successful US impact imaging mission, last of series

Following the success of its predecessor, Ranger 8 was targeted at a second mare site (the equatorial region of the Sea of Tranquillity was of particular interest to the Apollo planners) but this time close to the terminator to provide higher-contrast lighting conditions.

The Agena's 90sec injection burn from the 185km Earth parking orbit produced a lunar miss distance of 1,828km. A 59sec mid-course correction burn on February 18, 159,743km from Earth, brought Ranger 8 back to the aim point of 2.6°N/24.8°E in the Sea of Tranquillity.

Instead of turning the cameras on 13min 40sec before impact as originally planned, ground controllers commanded the first of 7,137 pictures from 23min out, where resolution was comparable to that of the best Earth-based photographs. In addition, a manoeuvre to align the cameras along the velocity vector (reducing smear) was cancelled, partly to allow greater area coverage and better continuity with Ranger 7's pictures but also because there had been an unexplained telemetry loss during the course correction and controllers wanted to avoid further manoeuvres. The new scenes proved to be very similar to those of July 1964. The impact point at 2.59°N/24.77°E was within 24km of that planned.

See Ranger 7 for spacecraft description.

The two successful Rangers had satisfied Apollo's mare survey requirements – showing them to be sufficiently smooth for manned landings – so Ranger 9 was released for the more scientifically interesting 112km-diameter Alphonsus crater (13.3°S/3.0°W) in the highlands, where current volcanic activity had been suspected for some time. Lighting conditions had to provide good surface feature contrast, and if the launch were delayed alternate sites increasingly further west would have to be chosen. Target for a March 22 launch would be Copernicus, moving to Kepler on the 23rd and Schroter's Valley on the 24/25th.

The 90sec Agena injection burn from the 185km Earth parking orbit was the most accurate of the series, leaving the uncorrected impact point just 640km north of Alphonsus. A 31sec course-correction burn at 1203 GMT on March 23 and 282,244km out from Earth added 65.3km/hr to the velocity. Camera warm-up began at 1340 GMT on March 24, 20min before impact, with the first of the 5,814 images being returned from an altitude of 2,100km. A terminal manoeuvre of the kind cancelled on Ranger 8 pointed the cameras directly along the

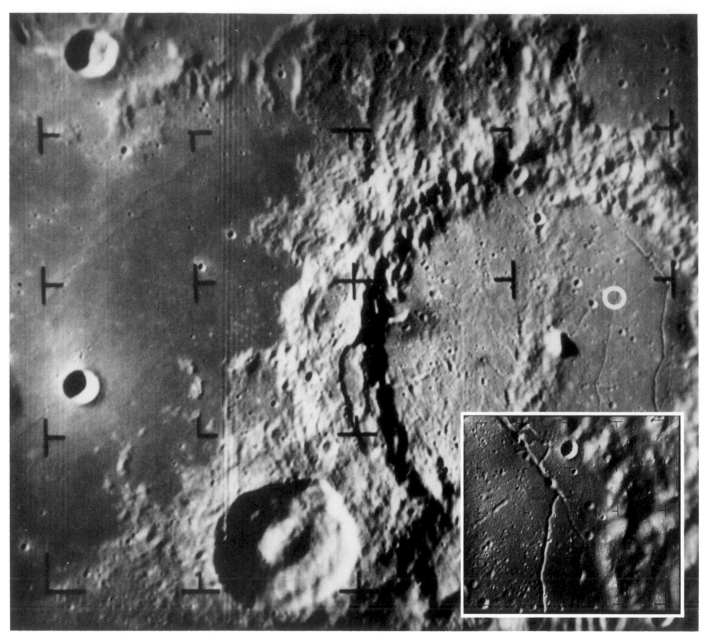

line of motion to reduce smearing. Resolution improved to 25cm just before impact at 9,617km/hr beside the 1,100m-high peak, only 6½km off the pre-mission target. The 14m-diameter impact crater was imaged by Apollo 16's orbital cameras in 1972.

Dark-haloed craters visible in the pictures appeared to be of volcanic origin and, as at the mare sites, there was a profusion of small craters.

Ranger 9 brought the $267 million (1965 values; $700 million in 1984 terms) programme to an end, its results having been perhaps more important to the engineers than to the lunar geologists or Apollo planners. Ranger tracking data were however used to refine the figure for the Moon's mass by an order of magnitude, and – an important finding – the centre of mass was discovered to be displaced from the geometric centre. The surface appeared to be covered in a layer of uncompacted material, but an answer to the question of whether it was deep enough to create a hazard to manned spacecraft would have to wait for the Surveyor and Luna landers.

See Ranger 7 entry for spacecraft description.

Kosmos 60 Launched March 1, 1965, by A-2-e from Tyuratam but failed to leave Earth orbit. Probably Luna 9-type lander. Preceding one-year hiatus in Soviet lunar flights was possibly due to launcher modifications following a string of failures.

Luna 5 Mass 1,476kg, launched May 9, 1965, by A-2-e from Tyuratam into 151 × 217km, 64.8° parking orbit. Acknowledged as lunar landing attempt. Course correction on May 10 resulted in impact at 31°S/8°W at 1910 GMT May 12 after the retromotor failed to fire.

Luna 6 Mass 1,442kg, launched June 8, 1965, by A-2-e from Tyuratam into 167 × 246km, 64.8° parking orbit. Acknowledged as lunar landing attempt. Main engine failed to cut off on time during course-correction manoeuvre of June 9, resulting in 160,000km flyby on June 11.

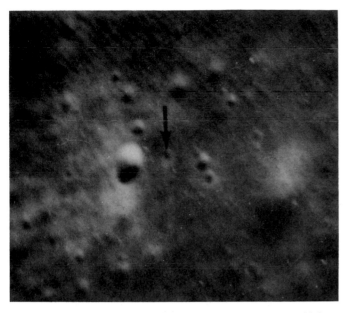

Left: *Ranger 9 rushes towards destruction in the large crater Alphonsus. North is at top and the picture covers an area about 200km on a side. The adjacent image, centred to the right of the white circle in the first, demonstrates the increasing resolution.*

Above: *This greatly enlarged image from Apollo 16 in 1972 located the Ranger 9 impact crater.*

Zond 3

Launched: July 18, 1965
Vehicle: A-2-e
Site: Tyuratam
Spacecraft mass: 1,145kg
Destination: Moon
Mission: Flyby
Arrival: July 20, 1965, closest approach 9,220km
Payload: Imaging system
Ultra-violet spectrograph (2,500–3,500Å)
UV and infra-red spectrophotometer (1,900–2,700Å, 3–4μm)
Meteoroid detectors
Radiation sensors (cosmic rays, solar wind)
Magnetometer
Ion thruster test
Radiotelescope
End of mission: March 1966
Notes: Imaged unseen remaining 30% of lunar far side

Early lunar and planetary mission attempts were marked by numerous failures. The Zond 2 Mars craft was launched late in the November 1964 window, possibly because of technical problems, and its Zond 3 twin was stood down. Rather than waiting 25 months for the next window, the Soviets used it for a Mars-duration voyage with a lunar flyby on the way out to exercise the photographic system. A successful flight would raise confidence for the imminent Venera 2/3 missions, using almost identical vehicles.

Zond was based around a 1m-diameter pressurised cylindrical compartment containing the electronics and batteries for communications and command. Solar panels on either side fed the batteries, although the arrays were not necessary for a Moon-only flight. Each panel carried a hemispherical radiator designed to lose heat transferred from the pressurised section by fluid. At the bottom of the cylinder was the experiment section. A landing capsule or the housing for the imaging system was accommodated here on Zond 3 and related craft. At the far end was the single KDU-414, 1.96kN correction motor. Three-axis attitude control was maintained by nitrogen gas jets responding to information from Sun and Canopus sensors; Zonds 2 and 3 also carried six ion thrusters to evaluate their potential use on future probes. The 3.5m-tall craft was completed by the 2m-diameter high-gain antenna.

Zond 3. **Key: 1** *magnetometer* **2** *engine section* **3** *guidance sensors* **4** *antenna* **5** *photographic section* **6** *hemispherical radiator* **7** *solar panels* **8** *high-gain antenna* **9** *pressurised compartment.*

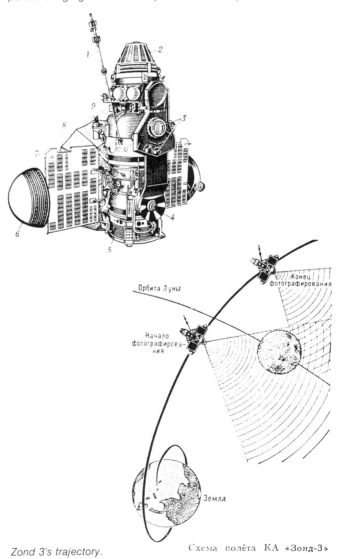

Zond 3's trajectory.

Схема полёта КА «Зонд-3»

Above: *Zond 3 imaged most of the lunar far-side area hidden from Luna 3.*

Below: *Venera 2.* **Key:** *1 engine section* **2** *magnetometer* **3** *hemispherical radiators* **4** *solar panels* **5** *scientific instrument section* **6** *attitude-control system tanks* **7** *attitude-control thrusters* **8** *pressurised compartment* **9** *high-gain antenna* **10** *antenna* **11** *solar sensor* **12** *solar and stellar sensors for attitude control.*

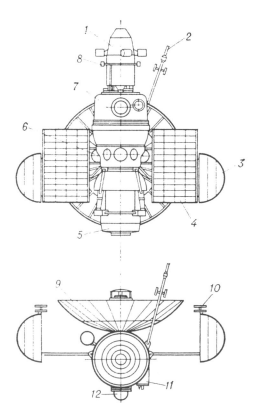

A 164 × 210km, 64.78° Earth parking orbit was followed by a final burn to Earth escape speed that took Zond 3 to the Moon in only 33hr. As the craft approached the Moon's illuminated leading edge, the imaging sequence began at 0124 GMT on July 20 at a distance of 11,570km from the near side. The camera system was similar in pinciple to that carried by Luna 3 in 1959. Images from an f/8, 106.4mm-focal-length lens system was focused onto 25.4mm-wide film that was later developed, fixed and dried for scanning by a fine light beam. Exposures were made at 1/30th and 1/100th of a second. A coarse scan at 67 lines/picture took just 135sec but the full-resolution 1,100 lines (860 elements/line) required 34min per picture. An attached UV spectrograph focused an image for exposure on Frames 8, 9 and 10 (of little value for lunar studies but useful for investigating the atmosphere of Mars).

Zond 3 made its closest approach and continued outwards, viewing at an advantageous low Sun angle the 30% of the far side missed by Luna 3. The sequence of 25 visual and three UV images, exposed at 134sec intervals, ended at 0232 GMT, with Zond 3 9,960km over the far side.

Since Zond 3 was actually a planetary probe, the controllers had to wait until July 29, when it had travelled 2.2 million km from Earth, before the narrow high-gain beam could lock on to the home planet. The high-quality pictures streamed back, and since this flight was intended to demonstrate communications reliability at large distances, the exercise was repeated on October 23 from 31.5 million km away. Other scans may also have been executed.

A 50m/sec course correction was made using the main engine on September 19 at 12.5 million km from Earth, again as a demonstration. Zond 3 sailed out to the distance of Mars' orbit and contact was apparently lost in early March 1966 when the craft was some 153.5 million km away.

This was the final planetary probe of the Zond series. Though Zonds 5–8 flew around the Moon and some acquired excellent photographs, they were all part of the Soviet manned lunar development programme. They are thus not described in detail in this book.

Luna 7 Mass 1,506kg, launched October 4, 1965, by A-2-e from Tyuratam into 129 × 286km, 64.8° parking orbit. Acknowledged as lunar landing attempt. Course correction on October 5 but engine ignited too early, at 2158 GMT on October 7, and craft crashed 6min later near 9°N/49°W to the west of crater Kepler.

Venera 2 Mass 963kg, launched November 12, 1965, by A-2-e from Tyuratam for Type I Venus flyby with photographic payload at 24,000km on February 27, 1966. Contact lost shortly before encounter. Similar to Zond 3 craft.

Venera 3 Mass 960kg, launched November 16, 1965, by A-2-e from Tyuratam for Type I Venus atmospheric capsule mission. Course correction on December 26 ensured impact at 0656 GMT (calculated) March 1, 1966, but contact lost shortly before encounter. Not known if instrumented and sterilised 90cm, 337kg capsule with parachute separated. Similar to Zond 3 craft.

Kosmos 96 Mass ~960kg, launched November 23, 1965, by A-2-e from Tyuratam for Type I Venus atmospheric capsule (?) mission but failed to leave parking orbit. See Veneras 2 and 3.

Luna 8 Mass 1,552kg, launched December 3, 1965, by A-2-e from Tyuratam into 181 × 221km, 51.9° parking orbit. Acknowledged as lunar landing attempt. Following successful midcourse correction, retro ignited too late and craft impacted at 9.13°N/63.30°W at 2151 GMT on December 6.

Luna 9

Launched: 1142 GMT January 31, 1966
Vehicle: A-2-e
Site: Tyuratam
Spacecraft mass: 1,538kg at launch; ~100kg surface capsule
Destination: Moon
Mission: Lander
Arrival: February 3, 1966, landed 184530 GMT at 7.13°N/64.37°W
Payload: Lander:
 TV camera (1.5kg)
 SBM-10 radiation detector
End of mission: Last contact 2255 GMT February 6, 1966
Notes: First lunar landing, first surface TV pictures

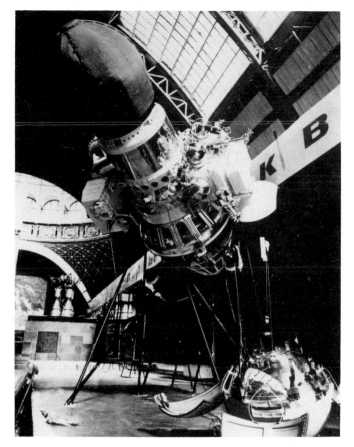

The Luna 9 landing capsule was housed in the spheroidal container at top, the side packages being ejected before retrofire. The altimeter dish is just visible at the rear. At bottom right is a replica of the lander itself.

The race for the Moon was very much on at the beginning of 1966. Activity in the US Apollo manned programme was reaching a peak, while there was little doubt that the Soviets had every intention of making manned attempts and were using unmanned probes to pave the way. It was also a matter of national pride to acquire the first images from the lunar surface.

The US Surveyor soft-landers were being prepared for their initial assault in late spring but the Soviets were first off the mark with Luna 9, the latest in a series of missions. The launcher placed it in a 167 × 219km, 51.9° parking orbit and before it had completed a full circuit of the Earth the escape ("e") stage ignited to establish a trans-lunar trajectory. This was only the second probe (Luna 8 being the first) to use the 51.9° orbit, which allowed an additional 130kg of payload.

Large numbers of planetary and lunar craft had previously failed to get beyond low Earth orbit, which might suggest an unreliable injection motor. In fact, as analyst David Woods has pointed out, if the whole vehicle was slow to orient itself for the injection burn only three attempts (in three orbits) were possible before the Earth's rotation took the craft out of contact with Soviet stations. When it hove into view the following day, its orbit would have decayed significantly through atmospheric drag and it would then be allowed to re-enter without any further attempt at ignition.

But Luna 9 departed successfully, separated from its final stage and began a 4°/sec "barbecue" roll to distribute the Sun's heat evenly. Luna was about 2.7m high at this stage and consisted of three main sections. At the bottom was a KTDU-5A liquid-propellant retromotor below a toroidal aluminium alloy tank containing an amine-based fuel; this was attached by a peripheral support framework to a 90cm spherical nitric acid oxidiser tank. Total propellant load was about 800kg. In addition, there were four arm-mounted thrusters drawing on the same supplies to provide stabilisation during the landing sequence; the US Surveyors incorporated a similar approach.

The central cylindrical section controlled the whole craft, carrying telecommunications and command units in a container pressurised to 1.2 Earth atmospheres. Strapped to either side were jettisonable packages totalling 300kg and pressurised at 0.13 atmospheres for other flight control functions. One housed the 93MHz radar altimeter which would trigger the final braking burn; it also held sets of nitrogen thrusters mounted on three arms and drawing on three spherical tanks for attitude control during the Earth–Moon journey.

The second unit supported Sun and Moon sensors for attitude reference, and both contained most of the craft's batteries; solar panels were not necessary for such a short mission. The topmost section contained the landing capsule, surrounded by a layer of thermal insulation.

Once tracking had provided sufficient data on the initial trajectory, the gas jets were used to kill the roll and orientate Luna 9 so that it first acquired the Sun and then the Moon before a 48sec course-correction burn was carried out at 1929 GMT on February 1, with the craft some 233,000km from the target. Shutdown was ordered by a pre-set onboard accelerometer. At 1316 GMT the following day, ground controllers established contact and checked the condition of the lander's TV system.

At a height of 8,300km and about an hour out, the gas jets again stopped the rolling motion to freeze Luna 9's alignment along the lunar vertical. This simple approach technique, while effective, limited Soviet landers to sites around 64°W near the equator, whereas US Surveyors could be directed to a wide variety of locations. Luna 9's trajectory meant that the landing system had no sideways motion to cancel out, only a 2.6km/sec vertical velocity.

Above: Part of the third panorama showing a 15cm stone some 1½m from the craft. At bottom one of the "petal" tips is visible, with an antenna at right.

Right: Soviet artist's impression of Luna 9 on the surface.

The radar triggered the terminal descent sequence as Luna 9 passed 75km altitude at 184442 GMT. The four stabilising thrusters burst into life, attitude reference was passed to an inertial (gyroscopic) system in the main control section, the two large side compartments were ejected to shed mass, and the main engine ignited to produce a thrust of 45.5kN. Five metres from the surface, a deployed sensor made contact and ordered shutdown as the capsule was thrown up and sideways, possibly by springs. The main bus impacted at up to 22km/hr.

The 58cm-diameter spheroidal capsule came to rest, its internal equipment protected by shock absorbers, and after 250sec a timer activated the radio transmitter: the first successful lunar landing had been achieved. It was not a true soft landing of the kind that Surveyor and the later Lunas would subsequently accomplish, but a survivable impact of the type intended for the failed US Rangers 3–5. Four spring-loaded "petals" opened to expose the top section and right the capsule. The interior was pressurised and fans forced circulation for cooling, as on earlier probes, but this time water was allowed to evaporate through a vent to help maintain the temperature within 19–30°C. Most of the equipment mass, including the batteries, was concentrated at the bottom to give a low centre of gravity. The upper half held the TV camera and radiation detector.

The first picture from the lunar surface was relayed to Earth during 0150-0337 GMT on February 4. The camera was of the simple facsimile type, drawing only 2½W and similar in principle to those used in the Viking Mars landers and the Soviets' own Mars capsules. A mirror could rotate through 360° to produce a 6,000-line panorama in 100min, although the view was limited to only 11° above and 18° below the normal horizontal. An 8cm-diameter rotating turret with the nodding mirror

protruded from the top, its height of 60cm above the surface making the horizon 1.5km distant. Small targets dangled from all of the antennae to provide brightness standards for calibrating the camera, and three thin vertical mirrors were spaced around the turret for limited surface stereo photography.

Signals at 183.538MHz carrying the first picture data were intercepted by England's Jodrell Bank radio observatory, where they were converted into images using a newspaper facsimile machine and released before the Soviets produced the official version. Luna 9 had landed at an angle of 16.5° in a relatively smooth area, although blocks several metres in diameter could be seen towards the horizon. Analysis of the terrain, with its depressions, mounds and rocks, indicated that the small craft had come to rest on the shallow slope of a 25m-diameter crater. The camera was in focus from 1.5m to infinity, with a best resolution of 1.5–2mm, about one third of the human eye's capability. But even the resulting rather murky images were proof positive that the lunar surface was neither fragile nor covered in a thick dust layer, and could support a craft of some size.

A second panorama, transmitted during the contact period at 1400–1654 later that day, revealed that Luna 9 had shifted to an angle of 22.5° to the horizontal, suggesting that it had landed on a stone or the soil was settling. This unexpected development shifted the camera position by several centimetres and permitted stereo photography.

Three further panoramas were received on February 5, and a session during 2037–2255 the following day produced images of smaller areas, bringing the total to nine full and partial scans by the time the batteries were exhausted. Total contact time was 8hr 5min over three days. The only other experiment reported a radiation dosage, possibly arising from the interaction of cosmic rays with surface material, of 30 millirads/day.

See Luna 13.

Kosmos 111 Launched March 1, 1966, by A-2-e from Tyuratam but failed to leave 191 × 226km, 51.9° Earth parking orbit. The flight of Luna 10 later in the month suggests that Kosmos 111 was an orbiting mission and not a lander of the Luna 9/13 type.

Luna 10

Launched: 1047 GMT March 31, 1966
Vehicle: A-2-e
Site: Tyuratam
Spacecraft mass: 1,582kg at launch (245kg in lunar orbit)
Destination: Moon
Mission: Orbiter
Arrival: April 3, 1966
Payload: Micrometeoroid detector (area 1.2m², for masses >7 × 10⁻⁸g)
Gamma-ray detector (0.3–4MeV)
Infra-red detector
Radiation detectors (solar wind, cosmic ray, X-ray)
End of mission: May 30, 1966 (after 460 revs)
Notes: First lunar orbiter

Just as the Americans originally intended to adapt their Surveyor lunar landers for an orbital role, the Soviets replaced the Luna 9/13 surface capsule with an ejectable satellite and used the standard Luna bus to brake it into a closed path around the Moon. Curiously, though, there was no imaging capability and the improvements introduced by Luna 12 suggest that Luna 10 was a stopgap solution to prevent the US Lunar Orbiter becoming the first artificial satellite of another world. Clark suggests it was based on a Kosmos-class Earth satellite.

A 200 × 250km, 51.9° Earth parking orbit preceded a course correction by the main engine on April 1 and a manoeuvre some 8,000km from the lunar surface to point the engine along the line of flight for the insertion burn. The motor ignited at 1844 GMT and cut 850m/sec off the approach speed

Luna 10 was ejected from a Luna 9-type main body. (Mike Ball)

to allow the Moon's gravity to pull the craft into a 350 × 1017km, 71.9°, 178min orbit. A landing mission requires a ΔV of about 2.6km/sec; the smaller burn needed by Luna 10 meant that less propellant (and therefore more payload) could be carried. It appears that the side packages ejected before retroburn to cut energy requirements, as on Lunas 9/13.

Twenty minutes later the 75cm-diameter, 1.5m-long chamfered cylinder was sent spinning away at about 2rpm. With only particles and fields experiments aboard, three-axis stabilisation was not necessary.

The orbiter was pressurised at slightly more than one Earth atmosphere and kept at a temperature of 24–26°C, probably by fan-induced circulation to transfer heat to the externally shiny shell for radiation to space. One of the satellite's first tasks was to beam a rendition of the *Internationale* back to Earth.

The magnetometer was deployed on a 1.5m boom following release and determined the magnetic field to be 10⁻⁵ the strength of Earth's at most. The gamma-ray detector recorded induced – by cosmic ray interactions – and natural radioactivity in the soil, showing it to be similar to terrestrial basalt. The

The separated Luna 10 in orbit.

radiation detectors did not uncover belts of trapped radiation or an ionosphere, but the detected cosmic ray background was higher than expected (five particles/cm²/sec). The 15 × 30mm plate of the IR instrument was used to measure heat flux from the Moon. The micrometeoroid detector was similar to those carried by Lunas 1–3 and showed impact rates to be greater than those encountered in interplanetary space. Orbital tracking provided a map of variations in the gravitational field that would be vital for subsequent accurate targeting of circling manned and unmanned landers before descent. Radio occultation measurements, looking for variations in signal strength when Luna 10 was close to the limb as viewed from Earth, were carried out in a vain search for a tenuous atmosphere.

With its batteries exhausted after 56 days of orbital operations and 219 communications sessions, the first lunar orbiter continued silently in a 378 × 985km, 72.03° path.

Luna-1966A Launched April 30 (?), 1966, by A-2-e from Tyuratam as lunar orbiter but failed to reach Earth orbit? Tentative evidence only.

Surveyor 1

Launched: 144101 GMT May 30, 1966
Vehicle: Atlas-Centaur 10 (Atlas No 290D)
Site: ETR 36A
Spacecraft mass: 995.0kg at launch (294.3kg on landing)
Destination: Moon
Mission: Soft landing
Arrival at target: Landed 061736 GMT June 2, 1966
Payload: TV camera
End of mission: January 7, 1967 (lost contact)
Notes: First lunar soft landing; second pictures from lunar surface

Although the Soviet Luna 9 transmitted the first images from the surface of the Moon, Surveyor 1 was the first true soft-lander. Surveyor was originally conceived by JPL as an extensive series of both landing and lunar-orbiting flights beginning in 1963, but ended up as seven soft-landers with a basic set of instruments. Despite the magnificent record of five successes

out of seven attempts, Surveyor did suffer severe development problems, and the total expected cost of $50 million eventually ballooned to $469 million ($2,000 million in 1984 terms).

Hughes Aircraft was awarded the spacecraft contract in early 1961; the other contenders were Space Technology Labs (later TRW), McDonnell Aircraft and North American Aviation. Surveyor evolved into a 1,040kg-class spacecraft (dictated by Atlas-Centaur throw capability) weighing about 283kg after touchdown. Surveyors 8–10, weighing 1,111kg each and carrying extra experiments, were planned but cancelled on December 13, 1966, because of the early successes.

Above: *Surveyor in landed configuration, with Leg 3 at left, 1 at right. The delicate electronics were housed in thermal compartments (seen here as a corrugated white and silver box) for protection against the harsh lunar night. Note the near-side landing radar and the crushable blocks of aluminium under the leg-body joints.*

Below: *Principal features of Surveyor 1/2. Adjacent to Leg 2 is the approach TV camera, carried by the first two craft but not used.*

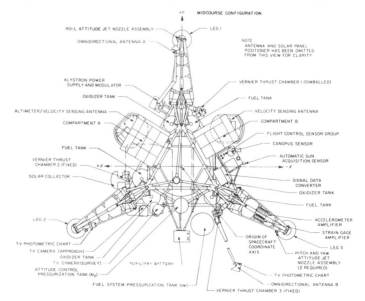

In landing configuration the 3m-high vehicle had a maximum width of 4.27m across its three legs. The basic framework was a 27kg thin-walled aluminium triangular structure with a leg at each corner and the large solid-propellant retromotor in the centre. Surveyor was topped by two panels, one carrying 3,960 solar cells to provide 85W of power to silver-zinc batteries and the other acting as the high-gain antenna to transmit the 600-line vidicon (TV) camera pictures to Earth. Every Surveyor carried this 7.3kg camera, with its vertically mounted optics scanning the surface to the 2½km-distant horizon via a mirror. The 200-line images were returned through a low-gain antenna.

The 11mm-square image on the vidicon surface took 1sec to scan for 600-line pictures, yielding a resolution of 1mm at a distance of 4m. The associated optics could be adjusted on command from Earth to f/4–f/22 aperture and 25–100mm focal length. Surveyor 1 carried a wheel with red, green and blue filters to permit colour images to be constructed; some later Surveyors used polarising filters instead. Surveyors 1 and 2 actually carried downward-viewing TV descent cameras to return Ranger-type images, but the requirement was deleted before the first launch and they remained unused. The sampler arm occupied that camera position on Surveyor 3.

Attitude control during the 60hr+ Earth–Moon journey was achieved through 0.27N nitrogen thrusters mounted on the legs. Mid-course corrections were made by three variable-thrust (133–483N) monomethyl hydrazine hydrate/nitrogen tetroxide thrusters, one of which could be swivelled to provide roll control.

At an altitude of about 96km a marking radar mounted inside the retromotor nozzle signalled that braking should begin. The 94cm-diameter, 35.6–44.5kN Thiokol motor was ignited 7sec later at about 76km altitude to cut the 9,600km/hr approach speed to about 400km/hr in a 42sec burn. On Surveyor 1 this motor accounted for 655kg of the 995kg launch weight. The three liquid engines were commanded by the autopilot to provide stability during this phase.

A radar altimeter Doppler velocity-sensing (RADVS) system provided the data for vehicle control once the solid motor had been ejected about 40km above the surface to clear the legs for landing. At this height radar returns from the ground were excellent and the RADVS fed the information into the computer for closed-loop control. Surveyor was an important step in proving this concept for the Apollo Lunar Module. The three liquid thrusters continued to fire to chop the descent speed to 5km/hr some 4.3m above the surface. To avoid disturbing the landing area with plume impingement, the thrusters were shut off and Surveyor 1 free-fell to an 11km/hr landing. The legs carried aircraft-type shock-absorbers, compressible footpads and blocks of crushable aluminium honeycomb under the leg-body joints. Strain gauges provided information on landing characteristics. The latter assumed particular importance once the programme became dedicated to Apollo. It would demonstrate the landing of a legged spacecraft on the Moon, with the scientific return regarded as secondary. The sites were selected largely for their Apollo potential.

Surveyor 1 was regarded as an engineering test when it was launched into a direct-ascent trajectory for a landing at 2.44°S/43.34°W in Oceanus Procellarum 63.6hr later. Sixteen hours out from Earth (0645 GMT May 31), the liquid thrusters performed a course manoeuvre to correct the 400km launching error. The marking radar gave its signal 95.49km above the lunar surface and the retro ignited at 75.22km with the craft travelling at 9,397km/hr. Surveyor 1 dropped onto the surface

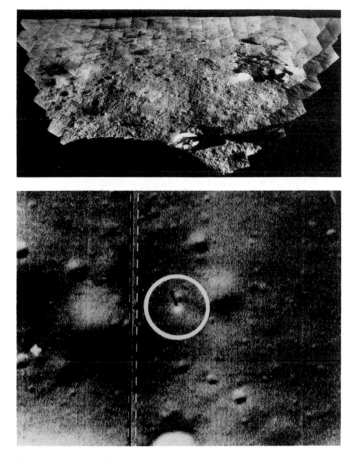

Top: *A montage of images from Surveyor 1, looking south-east with Footpad 2 visible. The crater at top right is about 3½m in diameter and 11m distant.*

Above: *Surveyor 1 (circled) was photographed on the surface by Lunar Orbiter 3.*

just 14km off target, at 2.45°S/43.22°W near the crater Flamsteed. The strain gauges showed that the three footpads made contact almost simultaneously, followed by a 6½cm, 1sec rebound caused by the 3m/sec vertical speed.

NASA scientists and engineers could hardly believe what they had done: a major first with an untried spacecraft after Ranger had required seven attempts to achieve its basic objectives. The first, 200-line, image, showing one of the footpads, was returned 36min after touchdown. Panoramic views revealed Surveyor to be in a 100km-diameter ghost crater flooded long ago by lava, with blocks more than 1m across scattered around. It was pinpointed on Lunar Orbiter 3 pictures and then found in Lunar Orbiter 1 and 5 images.

Surveyor 1 transmitted 10,732 images during the first lunar day and was put into hibernation as the Sun sank below the horizon. It survived the −160°C of the June 14–29 lunar night, although initial attempts to revive it on June 28 failed. Surveyor 1 eventually responded on July 6 and provided 618 further images. No more pictures were transmitted after the second night but contact remained possible until the following January.

Explorer 33 (Interplanetary Monitoring Platform D) Mass 93.4kg, launched July 1, 1966, by Delta No 39 from ETR 17A. Intended to enter 1,300 × 6,440km, 175°, 10hr lunar orbit for a six-experiment study of Earth's magnetic tail, interplanetary magnetic field and radiation. Four paddles with 6,144 solar cells provided power to control octagonal 71cm-diameter, 20cm-high body. Delta second stage produced excess velocity of 21.3m/sec, resulting in high Earth orbit of 15,897 × 435,330km. It was not possible to enter lunar orbit from this path, and the 37kg retromotor was fired to establish 30,550 × 449,174km, 28.9° geocentric orbit.

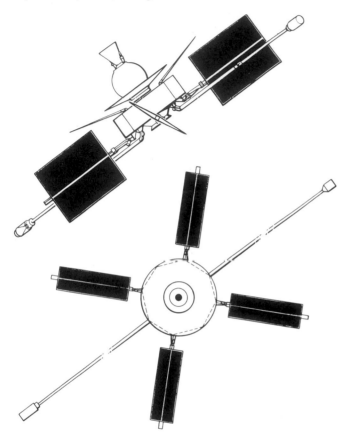

Explorer 33 was intended to enter lunar orbit but instead became established in high Earth orbit because of excess launcher velocity.

Lunar Orbiter 1

Launched: 1926 GMT August 10, 1966
Vehicle: Atlas-Agena D (Atlas No 5801)
Site: ETR 13
Spacecraft mass: 387kg at launch
Destination: Moon
Mission: Orbiter
Arrival: August 14, 1966
Payload: Photographic imaging system
Micrometeoroid detectors
Radiation dosimeters
End of mission: October 29, 1966 (lunar impact)
Notes: Second lunar orbiter (first US)

NASA's Langley Research Center issued a request for proposals for Lunar Orbiter on August 30, 1963, asking for an unmanned spacecraft that could return 1m-resolution images of the potential Apollo landing sites within 5° of the equator and between 45°E–45°W. Lockheed proposed a design drawing on its Agena-based spy satellites, but it was Boeing that received an $80 million contract for five spacecraft in the $200 million programme ($600 million in 1984 terms) on May 10, 1964.

The craft was three-axis-stabilised using Canopus and five Sun sensors and an inertial reference unit to control eight nitrogen gas thrusters. Four solar panels with a total of 10,856 cells provided 375W of power, with nickel-cadmium batteries used during periods in the Moon's shadow. The main engine for braking into lunar orbit and performing major manoeuvres was a 445N Apollo attitude thruster drawing on four propellant tanks carrying hypergolic (ie igniting on contact) nitrogen tetroxide and hydrazine. Mylar-dacron thermal blanketing maintained an internal temperature of 2–29°C.

The onboard flight programmer possessed a 128-word memory that allowed 16hr of automatic picture-taking operations, returning data through the 92cm-diameter 10W high-gain S-band (2.295GHz) antenna. When both the dish and its opposite omnidirectional antennae were deployed after launch, the 2m-high Lunar Orbiter measured 5.2m across, with the solar panels spanning 3.8m. Meteoroid detectors in a ring below the propellant tanks and two radiation dosimeters would provide data on the near-lunar environment. Thermal problems during the flight of Lunar Orbiter 1 forced a change in the white reflective paint for later missions.

The heart of the spacecraft was the 68kg Eastman-Kodak photographic system, encased in an ellipsoidal aluminium shell pressurised by nitrogen at 0.7–0.14kg/cm². It comprised

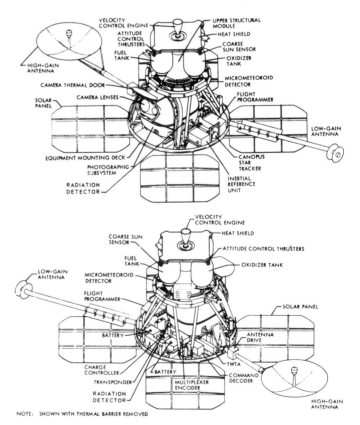

Principal features of Lunar Orbiter.

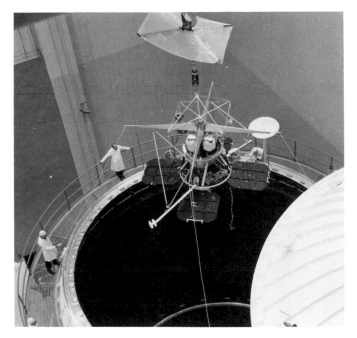

Above: Lunar Orbiter is lowered into a thermal/vacuum test chamber.

Above: The first image of Earth from the Moon was captured by Lunar Orbiter 1 at 1636 GMT August 23, 1966, from a lunar altitude of 1,198km. Lunar surface contrast is low because of the high Sun angle.

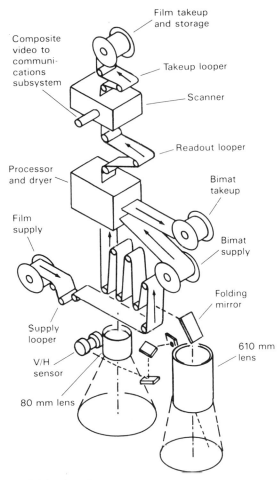

The Lunar Orbiter imaging system.

wide (80mm focal length) and narrow-angle (610mm) f/5.6 lenses viewing through a quartz window with an exterior mechanical thermal flap to produce simultaneous pairs of images on 70mm Kodak SO-243 high-definition aerial film at exposures of 1/25th, 1/50th and 1/100th of a second. The 79m of film allowed 212 image pairs to be captured, the film moving exactly 29.693cm between exposures. The exposed film was wound across storage spools, which could hold 21 frames, and into a processing unit where a layer impregnated with developing and fixing chemicals – essentially the method used in Polaroid instant cameras – was pressed against it for at least 3½min.

The negative image was scanned by a 0.005mm-diameter high-intensity light beam at 287 lines/mm and the signals generated from the photomultiplier beamed to Earth for image reconstruction. Each image pair took 43min to scan. Overlapping in the medium-resolution (80mm) pictures allowed stereoscopic mapping. Lunar Orbiter is the only US planetary craft to have used this imaging method; a vidicon TV system as carried by Ranger and Surveyor would have been far simpler but would not have produced the required resolution. A velocity/height sensor, an electro-optical device viewing through the 610mm lens, was used to move the film during exposure to match the image motion resulting from Lunar Orbiter's low altitude.

Lunar Orbiter 1 was taken into a 160km parking orbit by the Agena second stage, which followed a 28min coast with a 10min burn of its main engine. The spacecraft separated and unfolded its antenna and solar panels, but the Canopus star sensor, failed to lock on to its target; the Moon was used as a roll-control reference instead. A course-correction burn at 0028 GMT on August 11 set up Lunar Orbiter 1 for a lunar arrival within 80km of the planned point.

The primary mission was to photograph nine potential Apollo landing sites, seven secondary areas and the Surveyor 1 landing site to help pinpoint that vehicle, and to return the first clear images of the far side.

The first US craft to enter lunar orbit arrived on August 24 after a 92hr journey, with a 578.7sec, 789.65m/s burn establishing a 191 × 1,854km path inclined at 12.2° to the lunar equator and with a period of 3hr 37min. Tracking revealed that this orbit was being altered by variations in the gravitational field, a factor that had to be taken into account for the Apollo landings. In a major contribution to lunar science, the lava-filled near-side basins were eventually pinpointed as the cause (see Apollo 15 sub-satelite).

Engineering images were produced during August 18–20, and on August 21 a burn was carried out to reduce perigee to 58km in readiness for the mission proper; by August 25 the perigee had been reduced to 50km. The scientists and engineers were delighted to receive lunar surface pictures, but the high-resolution versions were blurred because the smearing compensation had completely failed. The medium-resolution images were excellent, however, and the last of the 211 frames was used on August 30. On August 23 (Revolution 16) the vehicle had captured the first Earth picture taken from lunar orbit; a second was recorded on Rev 26.

A press conference at NASA's Langley Research Center on October 6, 1966, was told that the sight of large blocks on the surface suggested that the Moon was strong enough to support a manned landing. The Ocean of Storms in the area of Surveyor 1 appeared to be the best landing site because it possessed 20% fewer craters than the others covered. The far side, seen previously only by Luna 3 and Zond 3, lacked the mare but there did appear to be lava-filled craters such as Tsiolkovsky. No meteoroid impacts were detected during the eight weeks spent in lunar orbit, whereas a strike every two weeks would be the norm in Earth orbit.

Lunar Orbiter 1, its film supply exhausted, fired its main engine for 97sec during Rev 577 to crash on the far side at 6.7°N/162°E at 1330 GMT on October 29. This was done to prevent its transmissions from interfering with those of the imminent Lunar Orbiter 2, and made it the first NASA probe to be deliberately destroyed.

Luna 11

Launched: 0803 GMT August 24, 1966
Vehicle: A-2-e
Site: Tyuratam
Spacecraft mass: 1,640kg at launch, ~1,100kg in lunar orbit
Destination: Moon
Mission: Orbiter
Arrival: August 27, 1966
Payload: As Luna 10 or 12?
End of mission: 0203 GMT October 1, 1966, after 277 revs

Luna 11 remains something of a mystery mission. Soviet reluctance to release photographs of the vehicle has prompted speculation that it was a failed imaging flight in the Luna 12 class. It appears that the payload and insertion bus remained as a single unit in lunar orbit, as did Luna 12, instead of splitting into two, as occurred with Luna 10. However, a Tass announcement claimed that Luna 11 would test spacecraft systems in lunar orbit as well as continuing the scientific investigation of near-lunar space, whereas the Luna 12 announcement specifically mentioned a photographic role. It is possible that a Luna 10-class vehicle was kept in one piece partly to give Soviet controllers experience in commanding the imminent imaging mission.

Launch was followed by a 193 × 201km, 51.8° parking orbit, and a course correction at 1902 GMT on August 26. A braking burn of about 900m/sec beginning at 2149 GMT on August 27 resulted in a 163.5 × 1,193.6km, 27°, ~3hr lunar orbit. A total of 137 communications sessions with Earth were held before battery exhaustion brought the mission to a close.

See Luna 10.

Surveyor 2 Launched 1232 GMT September 20, 1966, by Atlas-Centaur 7 from ETR 36A. Carried TV camera for soft landing in Sinus Medii but one of three thrusters failed to ignite for 9.8sec course correction at 0500 GMT on September 21, resulting in a 60rpm tumble. Despite 39 further attempts, the third thruster failed to start. Contact lost 30sec after retro ignition at 0934 on September 22, and spacecraft crashed at 5.5°N/12°W south-east of Copernicus. See Surveyor 1 for spacecraft and payload description.

Luna 12

Launched: October 22, 1966
Vehicle: A-2-e
Site: Tyuratam
Spacecraft mass: 1,620kg at launch, 1,136kg in lunar orbit
Destination: Moon
Mission: Orbiter
Arrival: October 25, 1966
Payload: Imaging system
Gamma-ray detector
Magnetometer
Radiation detectors (solar wind, cosmic rays, X-rays)
Infra-red radiometer
Meteoroid detectors
End of mission: Last contact January 19, 1967, after 602 revs
Notes: First images from Soviet lunar orbiter

Luna 12 was acknowledged by the Soviets as an imaging mission. The craft was based on the standard Luna bus but carried a larger payload section with a conical radiator to lose excess heat. An imaging system of the type carried by Zond 3 was housed in a separate package located where the ejectable radar altimeter resided on landing missions.

An Earth parking orbit of 199 × 212km, 51.9° was followed by a course correction on October 23 and a standard Luna approach to the Moon. On October 25, beginning 1,290km out, the main engine was ignited for 28sec to cut 937m/sec from the 2,085km/sec approach speed. The initial announcement gave an "equatorial" orbit of 100 × 1,740km with a 3½hr period; the figures were subsequently amended to 133 × 1,200km, 10° inclination. The vehicle remained as a single unit, with three-axis control using nitrogen gas jets and attitude information from the sensors in one of the side packages. The equatorial path suggests that Luna 12 was intended to image candidate manned landing sites, as Lunar Orbiter 1 had done shortly before in support of Project Apollo. The first imaging transmissions were not carried out until October 29, prompting speculation by Western analysts about initial control and pointing difficulties.

The picture-taking portion of the mission was completed early on, Luna 12's lifetime being limited by the battery supplies. The camera package was activated close to perigee, where a resolution of 15–20m was said to be possible. The film was developed, fixed and dried automatically and then scanned at 67 lines/frame for transmission to Earth at 135sec/frame. These quick-look images were followed by a full 34min scanning of each at 1,100 lines/frame (860 pixels/line). Examples were displayed on Soviet TV but few have ever been released, possibly because they suffer in comparison with Lunar Orbiter's or would reveal the areas being studied for the manned programme. The number obtained would have been limited by the film supply.

Once the primary mission was completed, Luna 12 was converted into a particles and fields satellite of the Luna 10 type, turning around its roll axis to record radiation from the lunar surface and deep space, and detecting meteoroid impacts. It was later revealed that space tests on electric motors to power the Lunokhod Moon rover were carried out by Lunas 12 and 14. The mission ended on January 19, 1967, when the batteries became exhausted after 302 periods of radio contact with Earth. See Lunas 9, 10, 11 and 14.

Above: Luna 12.
1 Pressurisation spheres *2* Camera package
3 Thermal control radiation *4* Radiometer
5 Instrument compartment *6* Batteries
7 Orientation system package *8* Antenna
9 Orientation system electronics *10* Altitude-control engines
11 Main engine

Below: The lunar surface south-east of crater Aristarchus, as seen by Luna 12. The smallest features are 15–20m across.

Lunar Orbiter 2

Launched: 2321 GMT November 6, 1966
Vehicle: Atlas-Agena D (Atlas No 5802)
Site: ETR 13
Spacecraft mass: 390kg at launch
Destination: Moon
Mission: Orbiter
Arrival: November 10, 1966
Payload: As Lunar Orbiter 1
End of mission: October 11, 1967 (lunar impact)
Notes: Fifth orbiter, first to make plane change

Whereas Lunar Orbiter 1 photographed potential Apollo landing sites in the southern region of the near-side equatorial area, its successor was targeted to capture 13 primary and 17 secondary candidate sites in the northern half. The Agena D and its payload first entered a 160km parking orbit for a 14min coast before departing for the Moon. Lunar Orbiter's 445N main engine was fired on November 8 for an 82km/hr ΔV burn some 265,485km out from Earth. Two days later, at 2056 GMT, the engine flared into life again some 2,027km from the Moon to brake Lunar Orbiter 2 into a 196 × 1,871km, 12.2°, 3hr 38min initial orbit. Another burn at 2258 on November 15 established a perigee of 50.5km in readiness for the main mission to begin on November 18.

By November 25, 208 of the planned 211 pictures had been taken – they included large areas of the far side – and the craft had made a total of 205 attitude changes. Three micrometeoroids had been recorded by the 20 thin-walled pressurised containers.

Readout of the images ended a day early on December 6 when the high-gain transmitter failed and three medium and two high-resolution pictures of Apollo site 1 were lost (earlier coverage was adequate, however). The scene looking across crater Copernicus from a height of only 45km is perhaps the best remembered of the entire Lunar Orbiter series: for the first

time the Moon could be seen as a three-dimensional body. Newspapers ran it as "the picture of the century".

On December 8, the photographic mission over, the engine was fired for 62sec to change the orbital plane to 17.5° to yield tracking data over a greater portion of the gravitational field and provide ground controllers with experience of working in higher-inclination orbits (the final two Orbiters would circle over the poles). A 3sec burn on April 14, 1967, shortened the orbital period by 65sec to reduce the time spent in darkness during the eclipse of April 24. A braking burn of 257km/hr deliberately crashed Lunar Orbiter at an estimated 4°S/98°E on the far side on October 11, 1967, to prevent interference with later missions.

"Picture of the century": an oblique view northwards across Copernicus from Lunar-Orbiter 2 at a height of 45.9km. The central peaks, rising 400m above the crater floor, were proposed in 1969 as an objective for the subsequently cancelled Apollo 20.

Payload: Lander:
        TV camera (1.3kg)
        Infra-red radiometer
        Penetrometer
        Radiation densitometer
        Radiation detector
End of mission: December 30, 1966, batteries exhausted
Notes: Second Soviet lunar lander, third overall

Luna 13

Launched: 1017 GMT December 21, 1966
Vehicle: A-2-e
Site: Tyuratam
Spacecraft mass: 1,620kg at launch, 112kg surface capsule
Destination: Moon
Mission: Lander
Arrival: Landed 1801 GMT December 22, 1966, 18.57°N/60.00°W

The Soviets closed 1966 in much the same manner as they had opened it – by rough-landing a small instrumented capsule on the Moon. Following a 171 × 223km, 51.8° Earth parking orbit and a course correction on December 22, the main retromotor was ignited at 1757 GMT on Christmas Eve just 70km above the Ocean of Storms some 400km from the Luna 9 site. Less than a second before the main craft crashed at 30km/hr, the capsule was ejected and four minutes later the covering petals had opened and transmissions begun.

Top: Luna 13 photographed its own shadow after landing, showing the double-turret arrangement of camera and radiation detector.

Above: The Luna 13 capsule. At rear is the penetrometer, in the foreground is the radiation density meter. The camera is the right-hand turret.

Unlike Luna 9, the new craft carried a three-axis accelerometer inside its pressurised body to record the landing forces for an investigation of soil structure down to a depth of 20–30cm. A further addition was a pair of 1.5m-long spring-loaded booms that deployed as the petals opened. One carried a titanium-tipped 5cm-long, 3.5cm-diameter rod that was pushed into the ground with a force of 70N by a small explosive charge at 1806 GMT. A penetration of 4½cm indicated a granular mixture similar to medium-density terrestrial soil, with a density of ~0.8g/cc. The second boom's experiment produced a similar result down to a depth of 15cm. A head containing a caesium-137 gamma-ray source was held against the surface and the degree of scattering by the soil measured by three detectors at increasing distances.

Four radiometers around the capsule's circumference picked up infra-red radiation from the surface that indicated a noon temperature of 117 ±3°C. A sensor mounted adjacent to the camera turret showed that about a quarter of the incident particle radiation was reflected from the surface.

The first panorama, scanned in a communications session beginning at 1353 on December 25, revealed only a few stones lying around and no major features. As with its predecessor, the area suggested lava flooding with erosion. The image also showed the craft to be tilted 16°. Four further panoramas were transmitted before the batteries were exhausted.

Luna 13 ended the first stage of Soviet lunar surface exploration by simple craft, allowing Surveyor to dominate the scene into 1968. The new series of heavy landers on the large Proton booster was not introduced until 1969, and the success of the Luna 16 sampler in September 1970 followed possibly six failures.

See Luna 9 for spacecraft and landing sequence description.

Lunar Orbiter 3

Launched: 0117 GMT February 5, 1967
Vehicle: Atlas-Agena D (Atlas No 5803)
Site: ETR 13
Spacecraft mass: 385kg at launch
Destination: Moon

Mission: Orbiter
Arrival: February 8, 1967
Payload: As Lunar Orbiter 1
End of mission: October 9, 1967 (lunar impact)
Notes: Sixth lunar orbiter (third US)

Lunar Orbiter 3 completed the mapping of potential landing sites for Apollo, freeing the final two missions for global and scientific surveys from polar orbits.

A 4.3sec course-correction burn by the 445N engine 37.7hr after launch was followed by a 542.5sec braking burn 54.9hr later to establish an initial orbit of 200 × 1,850km, 21°. A further firing four days later reduced the perigee to 48km in preparation for the main photographic mission of February 15–23.

Lunar Orbiter 3 produced 211 frames covering 12 promising Apollo and Surveyor sites identified during the first two missions. But only 182 had been returned to Earth when the motor winding the film on burned out on March 2 during Rev 149. Nevertheless, the Apollo requirements were satisfied and 15.5 million km² of the near side and 650,000km² of the far had been included for scientific purposes. A February 21 picture located Surveyor 1 on the surface. The images from the three Lunar Orbiter missions allowed eight preliminary Apollo targets to be selected by early April 1967. Site 2, at 0.75°N/24.2°E in the SW of the Sea of Tranquillity, received astronauts Neil Armstrong and Buzz Aldrin in the first manned landing, at 2018 GMT on July 20, 1969. The fifth area, where Surveyor 3 touched down, was visited by Apollo 12 in November 1969. The Lunar Orbiter results were vital to Apollo planning, and the near-100% success of the five-mission series was both unexpected and unmatched for that era. Lunar astronauts commented on the value of being able to learn lunar geography from the pictures.

Lunar Orbiter 3 changed its orbit on April 12, 1967, to avoid excessive periods in darkness during the April 24 eclipse, and on August 30 a 125.5sec burn circularised the orbit 160km high to simulate an Apollo trajectory. A 32sec burn (ΔV 190km/hr) deliberately crashed the vehicle at around 14.6°N/91.7°W on October 9, 1967.

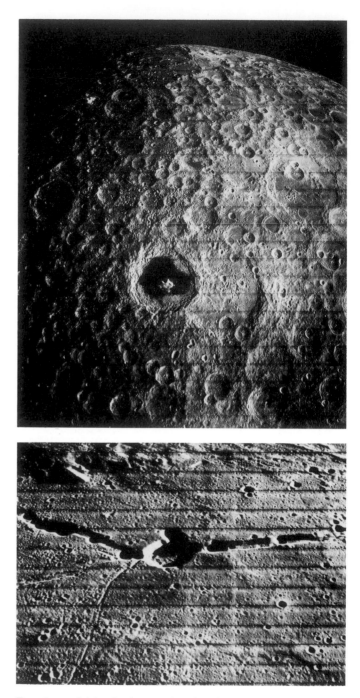

Top: Lunar Orbiter 3 picture taken from an altitude of 1,463km on February 19, 1967, features the distinctive lava-flooded, 250km-diameter Tsiolkovsky crater, discovered by Luna 3. The southern horizon at top reaches to within 650km of the south pole.

Above: Lunar Orbiter 3 was at an altitude of 62.8km and looking north across crater Hyginus (11km diameter) and its associated rille when this picture was taken. The area was a candidate for an Apollo manned landing.

Left: Taken from a height of 54.7km, this Lunar Orbiter 3 oblique view shows the 32km-diameter Kepler crater.

Surveyor 3

Launched: 070501 GMT April 17, 1967
Vehicle: Atlas-Centaur 12 (Atlas No 292D)
Site: ETR 36B
Spacecraft mass: 1,035kg at launch, 299kg on landing
Destination: Moon
Mission: Soft landing
Arrival: Landed 000417 GMT April 20, 1967, at 2.97°N/23.34°W
Payload: TV camera
Surface sampler
End of mission: May 4, 1967
Notes: Second US soft-lander (fourth successful lunar landing)

Surveyor 1's achievement in June 1986 was tempered somewhat by the failure of its successor that September. But Surveyor 3 produced another outstanding success in April 1967 and carried an expanded payload into the bargain. The Soil Mechanics Surface Sampler (SMSS) was to have been added to the later, more sophisticated, Surveyors, but programme managers decided to bring it forward to extend the capabilities of the early engineering missions. The SMSS used a simple tubular aluminium pantograph, powered by electric motors, to reach up to 1.5m from the lander. At the end was a 13cm-long 5cm-wide scoop capable of digging trenches up to ½m deep.

Surveyor 3 was the first of the series to be carried into an interim Earth parking orbit (166.5km circular) for a 1,329sec coast period before the Centaur engines demonstrated their new ability to re-ignite in space. A course correction by the three liquid thrusters 21hr 41min after launch targeted Surveyor about 370km south of Copernicus in the SE part of Oceanus Procellarum. Almost two days later, the braking motor ignited at a lunar approach speed of 9,451km/hr and burned out 40sec later, leaving Surveyor 3 descending at 502km/hr and the liquid thrusters firing to control the terminal

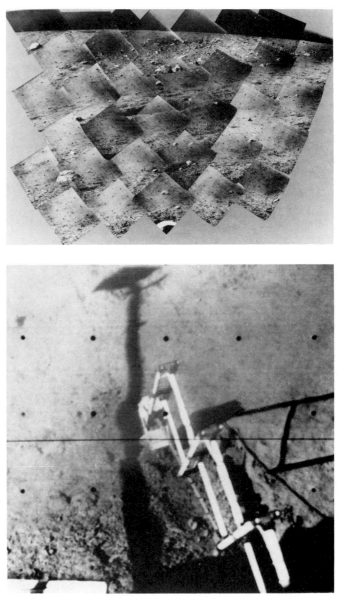

Top: *Surveyor 3's surroundings in the Ocean of Storms, looking north to the subdued crater wall.*

Above: *The sampler arm in operation on May 1, 1967.*

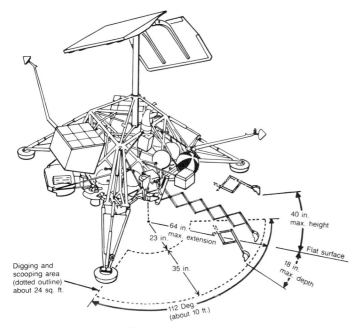

Operating range of the Surveyor scoop.

sequence. A few seconds before they should have cut out about 4.3m above the surface, the RADVS radar lost its lock (possibly because of reflections from large rocks scattered around the area) and guidance switched over to basic inertial control, allowing the thrusters to continue firing. They ensured that Surveyor 3 had the lowest landing speed of the entire programme, but an associated sideways motion of 3km/hr caused it to bounce 20m to one side, followed by a second hop of 11m. A ground command cut off the thrusters 34sec after initial contact but Surveyor still had enough energy to perform a final, 30cm sideways shuffle, coming to rest on a 14° slope. Some 6.8kg of propellant had been consumed between first and second contacts.

The multiple touchdowns fortuitously provided information on surface properties from studies of the footprints. But the thrusters may also have thrown up clouds of soil that partially coated the TV mirror and degraded the returned images. There were also difficulties in moving the mirror, suggesting that dust had penetrated its workings.

Surveyor 3 began transmitting the first pictures 58min after landing. These showed it to be near the rim of an old, subdued crater strewn with blocks up to 4m long – rather like the Surveyor 1 site, in fact. Matching the surface features with those on Lunar Orbiter 3 high-resolution frame No 154 allowed the site to be pinpointed.

A pyrotechnic locking pin was fired to release the arm under the control of JPL engineer Floyd Roberson. He moved it a step at a time, checking progress via TV while contending with the fact that the camera mirror reversed all the images. A total of 5,879 commands were sent in 18⅓hr of operations, resulting in the digging of four trenches, seven bearing tests and 13 impact tests. The bearing tests consisted of pressing on the regolith with the scoop's flat underside; in the impact tests the scoop was dropped on to the surface. Information was lost, though, because current telemetry from the arm's azimuth and elevation motors was too poor to allow the exerted forces to be

calculated. NASA's *Surveyor Programme Results* later reported that "The lunar surface material appears to have the properties and behaviour of a fine-grained terrestrial soil." The density did not appear unusual but the trenches revealed a crust of brittle material 2.5–5cm deep. Roberson was also able to pick up a small white rock in the scoop's jaws after 1½hr of manoeuvrings; a pressure of 7kg/cm² failed to break it.

Though Surveyor 3 did not survive the lunar night, it had already returned 6,326 TV pictures and executed a sophisticated series of arm movements. But this was not the end of its contributions to lunar science. Its surface position was accurately known and NASA took the opportunity to target the second manned lunar landing, Apollo 12, within walking distance of it. It was a demanding requirement in view of the fact that Apollo 11 had been more than 6km off target.

Mission commander Pete Conrad and pilot Alan Bean guided Lunar Module *Intrepid* down to a landing 160m NW of

Pete Conrad poses with his left hand on the extended sampler and his right on the TV camera of Surveyor 3. Lunar Module Intrepid, *accompanied by a dish-like S-band antenna, stands on the horizon.*

Surveyor on November 19, 1969. One of the first features Conrad saw after stepping on to the surface later that day was the old spacecraft resting on the other side of what had become known as Surveyor Crater. The two lunarnauts approached Surveyor from the SW during their second walk that same day. They photographed the trenches and pad prints, snipped off cable samples, removed the soil scoop and wrenched off the whole camera for return to Earth. Engineers wanted to see how well the materials had survived 31 months' exposure on the Moon's surface: the cabling had become brittle and the white paint had discoloured to tan. Tests also revealed one type of micro-organism, though that could have been deposited during the homeward journey.

See Surveyor 1 for basic spacecraft details.

Lunar Orbiter 4

Launched: 2225 GMT May 4, 1967
Vehicle: Atlas-Agena D (Atlas No 5804)
Site: ETR 13
Spacecraft mass: 390.1kg at launch
Destination: Moon
Mission: Orbiter
Arrival: May 8, 1967
Payload: As Lunar Orbiter 1
End of mission: July 17, 1967 (contact lost)
Notes: First lunar polar orbiter, seventh lunar orbiter (fourth US)

The first three missions having successfully photographed the potential Apollo landing sites, Lunar Orbiters 4 and 5 were released for scientific surveys of the Moon. A course correction of 53.8sec with the 445N engine at 1645 GMT on May 4 was followed by a 501.7sec, 659m/sec ΔV main braking burn at 2154 on May 8 into a 2,705 × 6,034km, 85.48° path. The high orbit and inclination allowed the cameras to record 99% of the near side and some of the far just over the limb down to resolutions of 60m in 163 frames (some were lost because of thermal and film fogging problems). The south pole was photographed for the first time. Film readout to Earth ended on June 1. The new thermal-protection requirements of the polar orbit were met by more than 500 quartz mirrors on the body's underside.

Lunar Orbiter 4's orbit was lowered on June 5 and 8 to 77 × 3,943km to match that of Lunar Orbiter 5 in order to provide gravitational data in support of that mission. Contact was lost on July 17 and the spacecraft is believed to have impacted in late October through gravitational perturbations.

Venera 4

Launched: 0240 GMT June 12, 1967
Vehicle: A-2-e
Spacecraft mass: 1,106kg at launch (383kg Venus capsule)
Destination: Venus
Mission: Atmospheric probe
Arrival: October 18, 1967 (impact 19°N/38° longitude)
Payload: Capsule:
    Radio altimeter
    Temperature and pressure sensors
    Gas analyser
    Accelerometers
  Bus:
    Magnetometer
    Cosmic ray detectors
    Ion detectors (Venus ionosphere)
    Ultra-violet photometer
End of mission: October 18, 1967
Notes: First successful penetration of Venusian atmosphere; first successful Soviet planetary mission

The dense clouds concealing Venus have made it one of the most difficult planets to study from spacecraft. Up to the late 1970s only general surface features around the equator were known, through Earth-based radar studies, and it was not until the advent of orbiting radar craft (see Veneras 15/16 and Pioneer Venus Orbiter) that maps covering large areas could be compiled. The Soviet's Venus achievements stand in stark contrast to their record at Mars: the first orbiter, atmospheric capsules, surface imaging and in situ analysis. However, at least ten failures had been caused by launch vehicle and communications problems before Venera 4 sent the first transmitting capsule slicing through the atmosphere. This

Left: Operating at an altitude of 2,723km in May 1967, Lunar Orbiter 4 revealed the spectacular Orientale impact basin on the lunar far side. The outermost ring of Cordillera Mountains is 900km in diameter.

Below: Venera 4. **Key: 1** *pressurised compartment* **2** *star sensor* **3** *Sun sensor* **4** *attitude-control gas bottles* **5** *Sun-Earth sensor* **6** *magnetometer* **7** *high-gain antenna* **8** *low-gain antenna* **9** *radiators* **10** *solar panels* **11** *vernier thrusters* **12** *attitude-control thrusters* **13** *cosmic particle detector* **14** *landing capsule.*

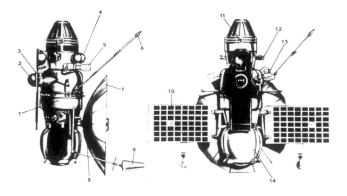

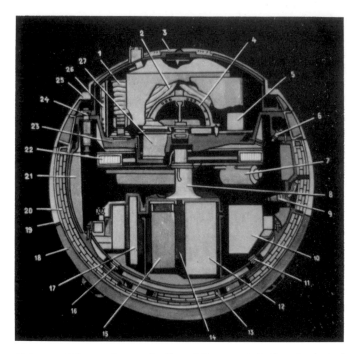

Cutaway of Venera 4 capsule. **Key: 1** *Drogue chute* **2** *main chute* **3** *pyrotechnic device for cap release (?)* **4** *antenna* **5** *pressure sensor (?)* **6** *valve (?)* **7** *de-humidifier* **8** *ventilator for temperature control* **9** *umbilical* **10** *commutator unit* **11** *accelerometers* **12** *transmitter* **13** *oscillation damper* **14** *batteries* **15** *transmitter* **16** *accelerometers* **17** *?* **18–21** *thermal protection* **22** *temperature-control system* **23** *top of capsule after cap release* **24** *cap-release device* **25** *cap* **26** *radio altimeter entrance* **27** *gas analyser.*

mission initiated a sequence of dual launches for every Venus window from 1967 to 1984 with the exception of 1973 and 1980.

Venera 4 was fired on a Type I trajectory to Venus from an Earth parking orbit of 173 × 212km, 51.78° (Soviet probes avoided the longer Type II paths), calculated planet miss distance being 60,000km. By July 17 it had travelled some 80 million km from Earth. The craft was similar to Venera 3, spanning 4m across the deployed 2.5m² solar panels and about 3½m high. A 2.3m-diameter high-gain antenna was employed for high-rate data; two conical antennae peering under the panels could also be used for this task. Three-axis control using star (Sirius was used in the subsequent Venera 5 and 6), Sun and Earth sensors was maintained by nitrogen jets.

The course correction on July 29, some 12 million km from home, was performed with the single KDU-414 1,960N unsymmetrical dimethyl hydrazine/nitric acid engine mounted at one end of the craft. Clark estimates that ~35kg of propellant was carried by each of Veneras 2–8. As in all Soviet probes, the cylindrical main body was pressurised and cooled to protect the electronics and batteries, and a shield covered the sunward side during cruise. A 2m magnetometer boom protruded from behind the main antenna.

Venera 4 approached the planet on a collison course on October 18 after a journey of 338 million km. The straps around the 1m capsule were released at 0434 GMT just before the main spacecraft hit the atmosphere and was destroyed. The new type of capsule, designed to withstand 350g decelerations, plunged in at 10.7km/sec near the equator and about 1,500km over into the night side. The thick layer of ablative heatshield glowed at 10,000–11,000°C from the fierce friction as it slowed. The battery – allowing a life of only 100min – and the other heavy equipment were arranged to provide a low centre of gravity, so presenting the greatest protection to the heat and leaving the parachute section at the top. The capsule was also designed to remain afloat and returning data should it splash down in an ocean.

The cap was jettisoned once a speed of 300m/sec had been reached, deploying a braking parachute in preparation for the main 'chute, both made of material designed to withstand up to 450°C. The scientific instruments were turned on at 0439 GMT as the rate of descent dropped to 10m/sec. Transmissions continued for 94min before the increasing pressure

probably broke through the thinner upper part of the pressure vessel. Soviet announcements claimed that transmissions began at a height of 26km and continued all the way down to the surface, where a pressure of 15–22 Earth atmospheres and a temperature of 280°C were indicated. But it emerged that the altimeter readings were open to misinterpretation and subsequent analysis – partly prompted by Mariner 5's contemporary readings – showed transmission beginning at 55km and ending 27km above the surface. These results indicate 500°C and 75 atmospheres at ground level.

Simple gas analysers sampled the atmosphere directly for the first time and produced a major surprise. Scientists had expected a mainly nitrogen atmosphere with little carbon dioxide (1–10%) but Venera found 90–95% carbon dioxide and no nitrogen. Oxygen and water vapour were present in small quantities.

The bus established a lower limit for the strength of the Venusian magnetic field of <1/5,000th that of Earth's, and detected a shock wave in the solar wind caused by the planet. There was no ring of trapped radiation *à la* the Earth's van Allen belts. The UV instrument found a weak atomic hydrogen corona 10,000km out but no atomic oxygen. See Veneras 5 and 6.

Mariner 5

Launched: 0601 GMT June 14, 1967
Vehicle: Atlas-Agena D (Atlas No 5401)
Site: ETR 12
Spacecraft mass: 244.8kg at launch
Destination: Venus
Mission: Flyby
Arrival: October 19, 1967, closest approach 4,100km at 173456 GMT
Payload: Ultra-violet photometer (4.18kg)
Solar plasma probe (3.12kg)
Magnetometer (3.65kg)
Trapped radiation detector (1.19kg)
End of mission: November 21, 1967 (formal mission); last telemetry December 4, 1967; partial contact October 14, 1968
Notes: Second US Venus flyby

Mariner 2 had completed a distant survey of Venus in 1962. NASA management approved the Mariner 5 project in December 1965 to penetrate ten times closer, making detection of the planet's magnetic field and its interaction with the solar wind more likely. Costs would be held down not only by flying a single spacecraft (Mariner 10 was the only other solo vehicle in the series) but also by modifying the Mariner 4

back-up vehicle. This bargain mission ran out at $100 million in 1984 terms, or 1/25th the cost of Viking.

Wherever possible, Mariner 4 equipment and instruments were used. But major changes were necessary because this craft had to head in towards the Sun instead of out to the colder Martian environment. The stubbier solar panels, 112 × 90cm with 17,640 cells to produce 370W at Earth and 550W at Venus, were reversed, with the cells underneath. As they deployed following separation from the Agena upper stage, the panels tugged on lanyards to unfurl a 1.226m², 25μm-thick aluminised teflon shield below the octagonal main body to provide protection against the 2.67kW/m² heat load at Venus and 4.16kW/m² at perihelion (Mariner 4's maximum had been only 1.43kW/m²). Otherwise, the thermal protection was the same as on Mariner 4.

The main course-correction engine, capable of two firings for a total ΔV of 92m/sec, and the attitude-control system, loaded with 2.35kg of nitrogen, were also the same, as was the command system. The tape recorder, with its 15m endless loop, had to store only science data at 66⅔ bits/sec for subsequent replay at 8⅓ bits/sec. The lack of image data meant that tape capacity could be cut by 80% from Mariner 4's 5.24Mbit.

The 53 × 117cm elliptical high-gain antenna was designed to operate in one of two positions, since the close trajectory would be strongly perturbed by the planet and post-encounter the Earth would appear in a very different position.

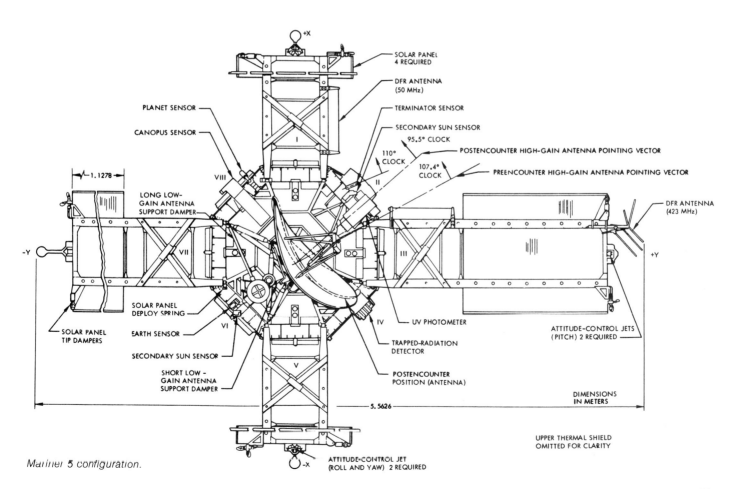

Mariner 5 configuration.

Three of the four instruments were modified versions of Mariner 4 designs; the fourth, the UV photometer, was new, having been deleted from the Mars mission. It would search for atomic hydrogen and oxygen emissions in the upper atmosphere. The TV camera on its scan platform had been removed.

The intention was to fly to within 3,218km of Venus on October 19, 1967, but when the Agena D's 10min burn freed Mariner from the 185km circular parking orbit over the south Atlantic the miss distance had been deliberately set at 75,000km. This was to ensure that neither vehicle would impact the planet and break quarantine. The first science data were received after Agena separation at 062717, and the spacecraft was set rolling for 16hr at 2 rev/hr in order to calibrate its magnetic field. Canopus was then acquired some 241,000km from Earth for three-axis control.

A 17.65sec course-correction burn at 230828 GMT on June 19 1.58 million km from Earth resulted in a ΔV of 15.39m/sec, 4½% less than expected but sufficient for a 4,100km flyby. The restart capability was not required.

The instruments continued to return cruise data until the 15.7hr encounter sequence was started by Earth command at 0249 on October 19, the stored information requiring 72½hr to play back from tape beginning the next day. The radiation team – headed by Dr James van Allen, who discovered the Earth's belts with Explorer 1 – found no similar particle radiation held about Venus by a magnetic field. Maximum strength of the field was put at 1% that of Earth's, and an upper boundary to the ionosphere was found at 500km (lowered to 350km at the time of Mariner 10, when the solar wind was more active).

The UV photometer picked up a hydrogen corona as strong as Earth's but no oxygen emission was detected; there was a faint UV glow over the night side. Occultation of the S-band signals as Mariner slid behind the planet for 26min was a major portion of the mission, producing temperature and pressure profiles through the atmosphere that extrapolated to 527°C and 75–100 Earth atmospheres at the surface. This was in complete disagreement with the recent, apparently in situ, Venera 4 measurements, and after considering the US results Soviet scientists reluctantly concluded that their capsule had not survived to landfall but had collapsed high in the atmosphere.

Mariner's close encounter over the night side and subsequent swing across the terminator towards the Sun altered its trajectory by 101.5° and allowed the planet's mass to be put at 81.50% of Earth's.

On November 21, with the mission at an end, JPL ground controllers switched communications to the omni antenna to put the craft into hibernation. Mariner 5 was now too distant for its transmission to be heard, but the intention was to reacquire it later to continue probing interplanetary conditions. Attempts began on April 26, 1968, and it took until October 14 to find the unmodulated carrier signal. No telemetry was received and commands met with no apparent response; the effort was terminated on November 5.

Kosmos 167 Mass ~1,100kg, launched June 17, 1967, by A-2-e from Tyuratam as twin craft of Venera 4. Failed to leave Earth parking orbit and re-entered after eight days.

Surveyor 4 Mass 1,037kg at launch, ~300kg on landing, launched 1153 GMT July 14, 1967, by Atlas-Centaur 11 from ETR 36A. Targeted for soft landing on lunar Sinus Medii with TV camera and surface sampler (see Surveyor 3). Contact lost at 0203 GMT July 17 2½min before landing; touchdown at 0.4°N/1.33°W could have been successful.

Above: *Explorer 35 is prepared for launch by Delta 50. Note the solid-propellant retromotor at top.*

Right: *This Lunar Orbiter 5 shot, taken from an altitude of 220km in August 1967, shows the prominent 85km-diameter ray crater of Tycho. Surveyor 7 landed later on the outer slopes and the crater was considered by scientists as a prime site for a manned Apollo landing. It was proposed in 1969 as an objective for Apollo 16 but the plan was eventually abandoned as too risky.*

Explorer 35 (Interplanetary Monitoring Platform 6) Mass 104kg at launch, launched July 19, 1967, by Delta No 50 from ETR 17B on direct lunar ascent with no means of course-correction. Similar to Explorer 33, it was designed to use Moon as anchor for probing interplanetary magnetic fields, plasma and micrometeoroid fluxes with seven instruments. As such, it is not described in detail here. The 36kg solid retromotor fired for 23sec on July 21 to brake the spacecraft into an 800 × 7,692km, 147°, 11½hr lunar orbit and was jettisoned 2hr later. Results showed that there is no lunar magnetosphere, solar wind particles impact directly on the surface (Apollo astronauts later collected samples on aluminium foil), and that the Moon creates a ''cavity'' in the solar wind stream.

Lunar Orbiter 5

Launched: 2233 GMT August 1, 1967
Vehicle: Atlas-Agena D (Atlas No 5805)
Site: ETR 13
Spacecraft mass: 390kg at launch
Destination: Moon
Mission: Orbiter
Payload: As Lunar Orbiter 1
End of mission: January 31, 1968 (lunar impact)
Notes: Eighth lunar orbiter (fifth US)

Lunar Orbiter 5 concluded a highly successful series by providing detailed coverage of 36 sites of scientific interest on the near side and mapping most of the far side from polar orbit. The remainder of the 213 frames included further images of five potential Apollo and several Surveyor sites.

The Canopus star tracker had difficulty locking on (but managed to do so 18hr 34min after launch) before the 26sec course correction on August 3. A 644m/sec, 508sec braking manoeuvre at 1648 GMT on August 5 produced a 196 × 6,040km, 85.0° lunar orbit, reduced two days later to 100 × 6,050km, 84.6°. Imaging began on August 6 and the final readout was made on August 27.

On October 10 a 41sec burn adjusted the orbit to 200 × 1,986km, 85.15° to reduce the period of lunar eclipse beginning on October 18. The spacecraft was used by Apollo tracking stations to practise for the manned lunar flights, and detailed orbital information allowed mapping of the Moon's gravitational field, revealing mass concentrations (mascons) associated with the near side lunar mare that could pull Apollo spacecraft out of position.

After an initial failure on January 18, 1968, the 155cm-diameter telescope of the University of Arizona's Planetary Laboratory photographed the spacecraft in orbit around the Moon on January 21, when it was turned to reflect sunlight off its mirrors and solar panels, appearing as a variable 12–15th-magnitude star. The main engine was fired at apogee for 18.9sec to cut speed by 103km/hr and allow the vehicle to

Lunar Orbiter 5 was photographed by the 155cm reflector of the University of Arizona on January 21, 1968, as the craft approached a lunar apogee of 2,022.5km. At bottom right is the over-exposed Moon, at top are two stars, and the spacecraft reveals itself as a short trail. The blobs have been added to mark the positions.

hit the surface around 0°N/70°W on January 31 after 1,200 revolutions.

The five highly successful Lunar Orbiter spacecraft – a sixth mission, carrying a gamma-ray instrument for the first time, was proposed – completed a total of 6,034 revolutions of the Moon and imaged 99% of the surface. This outstanding achievement, together with Surveyor, began an era of successful US space exploration.

Surveyor 5

Launched: 0757 GMT September 8, 1967
Vehicle: Atlas-Centaur 13 (Atlas No 5901C)
Site: ETR 36B
Spacecraft mass: 1,006kg at launch, 303.7kg at landing
Destination: Moon
Mission: Soft landing
Arrival: Landed 004642 GMT September 11, 1967, at 1.5°N/23.19°E
Payload: TV camera
Alpha-scattering instrument
Footpad magnet
End of mission: December 17, 1967, last data 0430 GMT
Notes: Fifth successful lunar lander (third US)

Surveyor 5 almost became the third failure in the series when a helium leak in the liquid thruster propellant pressurisation system soon after launch allowed the pressure to fall to 58.3kg/cm². At least 30.7kg/cm² was required to force the propellants into the combustion chambers, and the engines

The thrusters were reignited for 0.55sec on September 13 53hr after touchdown to disturb the surface: lumps of material were moved around and the alpha instrument shifted slightly but no craters were formed. By the end of the lunar day, on September 24, 18,006 TV pictures had been returned. Following its first lunar night Surveyor 5 responded immediately on October 15, relaying a further 1,048 images until sunset on October 24. None were recorded during the third day, but on the fourth a total of 64 200-line pictures concluded the work, overlapping the operations of Surveyor 6.

See Surveyor 1 for spacecraft description.

Surveyor 5 montage of lunar surroundings in mid-September 1967.

Surveyor 6

Launched: 1232 GMT November 7, 1967
Vehicle: Atlas-Centaur 14 (Atlas No 5902C)
Site: ETR 36B
Spacecraft mass: 1,008kg at launch, 300.5kg at landing
Destination: Moon
Mission: Soft landing
Arrival: Landed 010104 GMT November 10, 1967, at 0.53°N/1.4°W
Payload: As Surveyor 5
End of mission: December 14, 1967 (last data 1914 GMT)
Notes: Sixth successful lunar lander (fourth US)

Surveyor 6 was the third attempt at a soft landing in the Sinus Medii, the craft coming to rest on a level mare area near a ridge on top of a 10–20m-thick regolith (the greatest of the series). The alpha-scattering device revealed a surface similar to that found by Surveyor 5.

The Centaur had achieved a 90km miss distance but a course-correction 18hr 41min after launch had assured a precise landing.

On November 17, 177hr after landing, Surveyor 6 became the first spacecraft to be launched from the lunar surface when

would have ceased working far above the lunar surface had the normal descent profile been followed. Controllers rapidly designed a late braking sequence that required split-second precision for a safe landing. Instead of igniting the solid retromotor 83.5km up, with the liquid thrusters firing simultaneously for stabilisation, the burn was delayed for 12½sec to begin at 45.7km. This meant that the solid burned out at 1.34km instead of the usual 10.7km, at a speed of 96km/hr and close enough to the surface for the liquid thrusters to continue firing down to 4.3m.

The landing speed of 15km/hr was the highest of the entire series but still well within limits. Pre-landing thruster tests to investigate the pressurisation problem had pushed Surveyor 5 29km off target, and it came to rest just inside the inner 20° slope of a 9 × 12m rimless crater in the south-western Mare Tranquillitatis 25km NW of the Apollo 11 site. The angle caused it to slide gently, the footpads gouging out shallow trenches, and ruled out a full view of the surrounding terrain.

The alpha-scattering spectrometer was released on to the surface and the first in situ analysis of a celestial body showed the elemental composition to be similar to terrestrial basalt: more than half oxygen (probably in oxides), followed by silicon and aluminium.

Surveyor 6, essentially identical to Surveyor 5 but with a box hood on the camera.

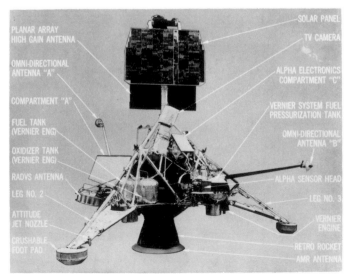

The Surveyor 6 landing site in Sinus Medii.

a 2½sec burn of its three thrusters, consuming 0.7kg of propellant, lifted it 3m high and a 7° tilt moved it 2½m to the west. The earlier landing marks could now be studied to determine surface mechanical properties, while the sideways translation provided a base for stereoscopic imaging (the TV cameras of the two final craft carried polarising filters instead of the earlier colour wheels). The alpha-scattering device came to rest upside down after returning 43hr of data.

The fourth successful US lunar lander returned 29,952 pictures up to a few hours after sunset on November 24. Hibernation began on November 26 and contact was regained only briefly on December 14.

See Surveyor 1 for spacecraft description.

Zond-1967A Launched November 21, 1967, by D-1-e from Tyuratam for lunar circumnavigation as part of manned lunar programme but failed to reach Earth orbit? Tentative identification only.

Surveyor 7

Launched: 0630 GMT January 7, 1968
Vehicle: Atlas-Centaur 15 (Atlas No 5903C)
Site: ETR 36A
Spacecraft mass: 1,040kg at launch, 306kg on landing
Destination: Moon
Mission: Soft landing
Arrival: Landed 010536 GMT January 10, 1968, at 40.86°S/11.47°W
Payload: TV camera + stereo mirrors

Alpha-scattering instrument
Surface sampler
Footpad magnet
End of mission: February 21, 1968 (last contact 0024 GMT)
Notes: Seventh successful lunar lander (fifth US)

The four successful Surveyors satisfied Apollo requirements in the Moon's equatorial zone, allowing Surveyor 7 to be released for a scientific mission. The ejecta blanket emanating from the bright, fresh ray crater Tycho in the far south was chosen, the rough highland region dictating a target area only 20km in diameter instead of the 60km of the earlier missions. For this reason two course corrections were planned, but the first was so accurate – leaving Surveyor 7 only 2½km off target – that no further alteration was necessary.

The three-legged vehicle touched down at 3.8m/sec some 29km north of Tycho's rim after a 66hr 35min flight. The camera revealed a rough area covered in blocks but, surprisingly, with fewer craters than the mare sites; there was a gentle slope of 3°. No other spacecraft has landed further from the equator, a planned late Apollo mission to Tycho never materialising.

After 20.9hr on the surface a pyrotechnic squib was fired on command from Earth to drop the alpha-scattering instrument to the surface, but the spectrometer stayed put. This was fortuitously the first flight with both a sample arm and a spectrometer, and the scoop was used to force the recalcitrant device to the ground. The arm later picked it up to analyse a rock and then the soil in a 1cm-deep trench, accumulating 63hr of data during the first lunar day. The main finding was a lower iron content than at the mare sites.

The scoop was used in 16 surface bearing-strength tests, dug seven trenches – one 40cm long and 15cm deep – and turned over a rock. One rock sample was "weighed" by lifting it and recording the required motor current; a value of 2.4–3.1g/cc was obtained. One rock was fractured, and on several occasions material stuck to the two magnets mounted on the scoop.

Some 20,993 pictures were recorded during the first lunar day and observations continued for 15hr after sunset at 0606 GMT on January 25. Images of the Earth and the Sun's corona out to 50 solar radii were obtained. Stereo imaging of small areas was possible by using a 9 × 24cm mirror mounted on the antenna mast, and on January 20 the TV had registered two 1W lasers aimed at the lander from observatories in California and Arizona. This demonstrated the feasibility of using lasers for communications and measuring the Earth-Moon distance with great accuracy (this was done later with laser reflectors left by Apollo and Lunokhod).

Surveyor 7 was reactivated at 1901 on February 12, but the long, cold lunar night had taken its toll and only another 45 200-line pictures were returned before it succumbed on February 21.

The $2000 million (1984 values) Surveyor programme – its cost was surpassed only by Viking on the planetary scene – yielded five outstanding successes from seven attempts. It produced more than 87,000 surface TV pictures, allowing Apollo to forge ahead with confidence, the spectre of a dangerous, dust-covered or sponge-like surface dispelled completely.

See Surveyor 1 for spacecraft description.

Zond 4 Launched March 2, 1968, by D-1-e from Tyuratam as test of Soyuz-based manned lunar craft to Moon distance (Moon was actually 180° away) followed by re-entry. Contact possibly lost and Zond entered heliocentric orbit or burned up in atmosphere (it is even rumoured to have landed intact in China). See Zonds 5–8.

Above: Surveyor 7's surroundings near crater Tycho. The distant hills and ravines are approximately 13km away. The debris at right centre sits in the 3m crater it created.

Below: The Zond 4-8 lunar craft were essentially Soyuz vehicles without the orbital module.

Luna 14

Launched: April 7, 1968
Vehicle: A-2-e
Site: Tyuratam
Spacecraft mass: ~1,600kg at launch
Destination: Moon
Mission: Orbiter
Arrival: April 10, 1968
Payload: Imaging system?
As Luna 12?
End of mission: June/July 1968?
Notes: Tenth lunar orbiter

As with Luna 11, few details of Luna 14 have been released. The lengthy gap following the successful flight of Luna 13 suggests that it could have been flown as a stopgap mission because of delays in introducing the new generation of heavy Luna samplers, rovers and orbiters launched by the Proton booster. It could even have been the Luna 12 back-up, brought out of storage to fly the last A-2-e lunar mission. No surface images were transmitted, but tracking in lunar orbit would have been useful for mapping the gravitational field in preparation for future missions requiring precise targeting.

A 189 × 242km, 51.8° Earth parking orbit was followed by a course-correction at 1927 GMT on April 8 using the main liquid-propellant engine. A 91 lm/sec burn to cut the approach speed of 2,190km/sec produced a 160 × 870km, 42°, 160min orbit. The date of battery exhaustion has not been revealed. See Luna 12.

Zond-1968A Launched April 22?, 1968, by D-1-e from Tyuratam as test of manned lunar craft but failed to reach Earth orbit. Tentative identification only. See Zonds 4–8.

Zond 5 Descent mass 2,046kg, launched September 14, 1968, by D-1-e from Tyuratam as test of manned lunar craft. Flew around Moon, coming to within 1,950km on September 18, and splashed down in Indian Ocean three days later after ballistic re-entry. Carried voice tape to test communications from lunar distance, and turtles and other biological samples for radiation studies. See Zonds 4–8.

Zond 6 Launched November 10, 1968, by D-1-e from Tyuratam as test of manned lunar craft with new re-entry method. Flew around Moon November 14, at 2,420km, made brief dip into Earth atmosphere to dissipate energy and re-entered for second time, undergoing lighter deceleration loads than those imposed by Zond 5 ballistic re-entry. Landed in USSR on November 17, carrying high-quality lunar photographs. See Zonds 4–8.

Venera 5

Launched: 0628 GMT January 5, 1969
Vehicle: A-2-e
Site: Tyuratam
Spacecraft mass: 1,130kg at launch (405kg capsule)
Destination: Venus
Mission: Atmospheric capsule
Arrival: May 16, 1969 (impact 3°S/18° longitude)
Payload: As Venera 4 + photometer in capsule
End of mission: May 16, 1969
Notes: Second successful Venus atmosphere capsule

The Venera 5 craft.

Venera 4's mission during the previous Venus window had demonstrated that the descent capsule was not rugged enough to survive all the way to the surface. The 19-month interval before the next launch opportunity was insufficient for the development of a more robust version, but the Soviets pressed ahead anyway using slightly improved craft that could withstand external pressures of 25–27 Earth atmospheres and decelerations of 450g. The main parachute, its fabric capable of withstanding up to 500°C, was reduced to two-thirds of its former area to produce a faster descent in the hope of penetrating further before the pressure vessel collapsed or heat knocked out the electronics. An improved radio altimeter was fitted in an effort to avoid a repeat of Venera 4's height confusion.

Venera 5 was injected into a Type I transfer path from its 203 × 218km, 51.82° Earth parking orbit with a 228sec final-stage burn over Africa at 0747 on January 5, producing a calculated Venus miss distance of 25,000km. On March 14 a 9.2m/sec correction burn by the large engine some 15½ million km from Earth targeted the bus and capsule for entries 2,700km into Venus' night side, close to the equator and at 20° longitude. The 1m sphere was ejected on May 16 some 37,000km and an hour out from the planet to begin an 11.18km/sec entry at 0601 GMT and 62–65° to the horizontal. The less favourable window meant the deceleration loads were about 50% greater than those suffered by Venera 4.

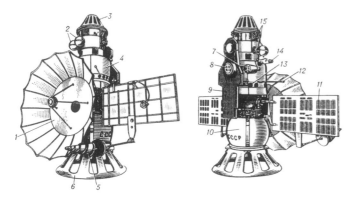

Venera 5. **Key: 1** *high-gain antenna* **2** *attitude-control gas bottles* **3** *engine section* **4** *pressurised compartment* **5** *low-gain antenna* **6** *handling equipment* **7** *solar sensor* **8** *star sensor* **9** *solar shield* **10** *descent capsule* **11** *solar panels* **12** *magnetometer* **13** *ion trap* **14** *Earth sensor* **15** *attitude-control thrusters.*

The 'chutes were deployed once air friction had cut the descent rate to 210m/sec, and the transmitter and instruments were turned on. The internal temperature crept slowly from 13°C to 28°C during the 36km fall to 24–26km, where the 27-atmospheres pressure (T = 320°C) ruptured the sphere and transmissions ceased after 53min. Based on those values, surface temperature and pressure were put at 530°C and 140 Earth atmospheres.

The combined gas analyser results from Veneras 5 and 6 indicated 93–97% carbon dioxide, oxygen less than 0.4% and 2–5% inert gas, in good agreement with Venera 4. The new photometer apparently took a single reading of a light flash 4min before destruction to reveal a light level of 25W/m²; the Soviets later dismissed this as probably a telemetry failure and noted that no data from the corresponding Venera 6 instrument were received. Water vapour at the 0.6 Earth atmosphere level was 4–11mg/L, indicating that the upper layers are not saturated.

See Venera 4, for spacecraft description, and Venera 6.

Venera 6

Launched: 0552 GMT January 10, 1969
Vehicle: A-2-e
Site: Tyuratam
Spacecraft mass: 1,130kg at launch (405kg capsule)
Destination: Venus
Mission: Atmospheric capsule
Arrival: May 17, 1969 (impact 5°S/23° longitude)
Payload: As Venera 5
End of mission: May 17, 1969
Notes: Third successful Venus atmospheric capsule

Venera 6 was launched shortly after its twin, via a 184 × 193km, 51.75° Earth parking orbit. But this time the booster was not so accurate, resulting in a Venus miss distance of 150,000km. A 37.4m/sec correction on March 16 some 15.7 million km from Earth targeted the spacecraft for

entry on the planet's night side. The capsule was separated 25,000km out from Venus on May 17 and entry, 300km from Venera 5's point the day before, began at 0605. Data were received for 51min before the pressure vessel collapsed at 26 Earth atmospheres and a height of 10–12km. The temperature and pressure values extrapolated to 400°C and 60 Earth atmospheres at the surface, rather at odds with its predecessor's findings. It was suggested that Venera 6 had landed on a mountain (leading the radio altimeter to give erroneous height values) or that the instruments had suffered some damage during the descent.

See Venera 4, for spacecraft description, and Venera 5.

Luna-1969A Tentatively identified Luna 16-type sample-return mission. Launched January 19, 1969, by D-1-e from Tyuratam but suffered booster failure and did not reach Earth orbit.

Mariner 6

Launched: 0129 GMT February 25, 1969
Vehicle: Atlas-Centaur 20 (Atlas No 5403C)
Site: ETR 36B
Spacecraft mass: 413kg at launch
Destination: Mars
Mission: Flyby
Arrival: July 31, 1969, closest approach 3,429km at 0519 GMT
Payload: TV system
Infra-red radiometer
Infra-red spectrometer
Ultra-violet spectrometer
(total instrument mass 59kg)
End of mission: July 31, 1969 (formal mission end)
Notes: Second successful Mars flyby (both US)

The two $148 million Mariner F and G flyby probes to Mars were approved by NASA on December 22, 1965, as more sophisticated follow-ups to the successful Mariner 4 of 1964–5. The more powerful Atlas-Centaur combination permitted a number of additional instruments, and an atmospheric entry probe was suggested but rejected on grounds of time

Mariner 6/7.

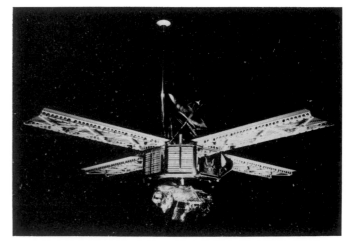

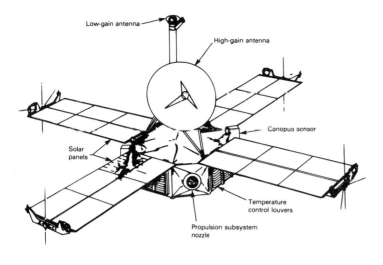

and cost. Narrow (508mm focal length) and wide (52mm) angle TV cameras would be carried on a scan platform along with other instruments designed to determine the main atmospheric constituents. The platform moved up to 70° in elevation and 215° in azimuth, controlled by a reprogrammable computer with a 128-word (22 bits each) memory.

A maximum data rate of 16.2kbits/sec (2,000 times that of Mariner 4) dictated a 20W, 102cm-diameter S-band high-gain antenna, the power being supplied from 17,472 solar cells on four 90 × 215cm panels (449W at Mars) and a 1,200W-hr, 14kg silver-zinc battery. A tape recorder could store 195Mbits of data for later playback to Earth. The deployed panels gave each Mariner a span of 5.79m.

The 91kg scan platform and its instruments were slung under the main octagonal magnesium body, 45.7cm deep and 138.4cm in diameter, with the antenna on top. This left the 225N hydrazine course-correction thruster, capable of two firings, protruding from one of the body faces; the others

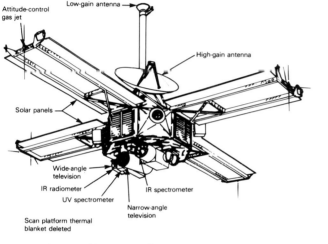

The Mariner Mars 1969 spacecraft.

Mariner 6 near-encounter images 17, 19, 21 and 23 (from left), taken at 84sec intervals. The 4,000km-long swath is about 725km wide and runs along 15°S in the Sabaeus Sinus region.

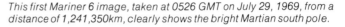

This first Mariner 6 image, taken at 0526 GMT on July 29, 1969, from a distance of 1,241,350km, clearly shows the bright Martian south pole.

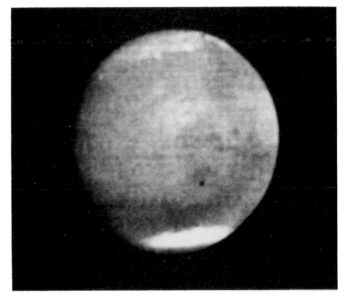

carried louvres for thermal control. Attitude control, using information from gyros and Sun and Canopus sensors, was provided by twin sets of six nitrogen thrusters mounted on the solar panel tips.

Mariner 6 flew on its twin's Atlas booster after its own had partially collapsed on the pad a few days before launch; the direct ascent into the Type I trajectory went as planned. A 5.35sec course correction on March 1 left the craft with a 3,540km miss distance at Mars.

The Canopus sensor was confused and lost lock after bright debris was shaken loose when a squib was fired to unlatch the scan platform. It was decided that at close encounter attitude control should be handled by the gyros to avoid the possibility of the instruments looking in the wrong direction.

On July 28, after 154 days of flight and with Mariner 6 still 1,255,000km from Mars, controllers at JPL ordered the scan platform and its complement of instruments to take a series of 33 images at 37min intervals. When they were played back later from the tape recorder over a 3hr period, scientists saw a turning globe growing visibly larger. The tape was erased and another 17 images were recorded before the actual encounter.

Close approach occurred just south of the equator over Meridiani Sinus and Sabaeus Sinus, 25 images of the after-noon and evening surface being stored on tape while scientific data were transmitted in real time. Unfortunately, one of the two infra-red spectrometer detectors had failed.

The TV images were replayed the next night and revealed heavily cratered lunar-like regions, chaotic terrain and desert-like flat areas. Equatorial surface temperature of $-73°C$ at night was recorded by the radiometer, with a low of $-125°C$ at the south polar cap. The long-range pictures of that pole showed an irregular border to the cap and scientists requested an increase in the number of Mariner 7 images from 25 to 33 in order to view this interesting region.

The cap appeared to be either solid carbon dioxide or covered by carbon dioxide clouds. Surface pressure was 6–7mb. Carbon dioxide was clearly the major atmospheric constituent (98%), whereas at the time of Mariner 4's flyby it had been assumed to be nitrogen. That gas could now account for only a few per cent at most.

A 38km-diameter crater exhibits slump terraces and radial gullies in this Mariner 6 images of an area at 5°E just north of the equator.

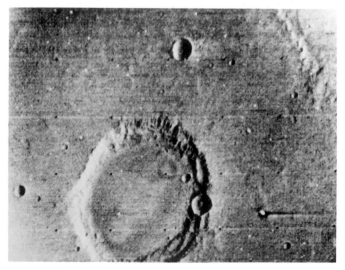

Mariner 7

Launched: 2222 GMT March 27, 1969
Vehicle: Atlas-Centaur 19 (Atlas No 5105C)
Site: ETR 36A
Spacecraft mass: 413kg
Destination: Mars
Mission: Flyby
Arrival: August 5, 1969, closest approach 3,430km at 050049 GMT
Payload: As Mariner 6
End of mission: August 5, 1969 (formal end)
Notes: Third successful Mars flyby (all US)

Mariner 7 demonstrated the advantages of carrying a reprogrammable command system after images from its pre-decessor revealed interesting detail around the south pole. Controllers were able to respond to the scientists' request that the 25 pictures planned for the near-encounter phase be increased to 33 to reach down to the pole as the scan platform slewed from north to south.

A 4m/sec, 7.6sec course correction on April 8 brought Mariner 7's aim point to within 190km of that desired and on May 8 it was placed under inertial (gyro) control in preparation for the unlatching of its scan platform in order to avoid a repetition of Mariner 6's attitude problems. It too would make the Mars encounter under gyro-referenced attitude control.

The craft intruded into the Mariner 6 close approach when the Johannesburg station lost signal for 7hr before a command to switch to the low-gain antenna brought a response within 11min. Fifteen telemetry channels had been lost, with others garbled, and tracking revealed that the point of closest planetary approach would be shifted 130km to the SE. Fortu-nately, all of the scientific instruments operated when they were turned on. Subsequent analysis showed that the single silver-zinc battery had failed just before signal loss, rupturing its case and spraying electrolyte out into space. The jet had acted like a thruster, altering the trajectory, and had also provided the medium for short circuits in some of the elec-tronics. Mariner 7 was fortunate to survive such a trauma. Positional calibration of the scan platform had been lost and distant Mars views were used to construct a new reference.

At 093233 on August 2 the first of 93 far-encounter images was recorded in a sequence stretching to 4½hr before closest approach. All of the instruments were operating smoothly and the 33 close pictures showed terrain similar to that uncovered by Mariner 6, except that the centre of Hellas was strangely devoid of craters down to the 300m resolution limit of the cameras. By contrast, the Hellespontus region just to the west was found to be heavily cratered. The edge of the south polar cap brought out the crater features in sharp outline. Atmos-pheric scattering from aerosols at 15–40km was clearly visible in some limb images. See Mariner 6 for spacecraft description and general mission results.

Mars-1969A and -1969B Mass 3,500kg?, launched March 27 and April 14, 1969, by D-1-e from Tyuratam but failed to reach Earth orbit. First of the second-generation Mars probes, with landing capsule for release by flyby bus. See Mars 2. Tentative identification.

Luna-1969B Tentatively identified Luna 16-type sample-return mis-sion launched June 4?, 1969, by D-1-e from Tyuratam but suffering booster failure.

Left: *The frost-covered Martian polar cap. The south pole is at top left.*

Below left: *Mariner 7 narrow-angle picture reveals frost-filled craters on the edge of the south polar cap.*

Luna 15 Mass 3,500kg?, launched 0255 GMT July 13, 1969, by D-1-e from Tyuratam. Course correction made on July 14, braked into lunar orbit by burn at 1000 GMT July 17 over far side. Orbit changes: 1308 GMT July 19, 96 × 221km, 126°, 123.5min; 1416 GMT July 20, 16 × 110km, 127°. While Apollo 11 *Eagle* was still on Moon, retromotor ignited 1547 GMT July 21 after 52 revs, spacecraft impacting 1551 GMT at 480km/hr at 17°N/60°E in Mare Crisium. Nature of mission unknown but generally considered to be Luna 16-type sample return.

Zond 7 Launched August 8, 1969, by D-1-e from Tyuratam as test of manned lunar craft. Flew around Moon August 11 at ~2,000km and made skipping re-entry August 14 to land in USSR, returning high-quality lunar photographs. See Zonds 4–8.

Kosmos 300 and 305 Launched September 23 and October 22, 1969, by D-1-e from Tyuratam but both failed to leave Earth orbit. Timings suggest they were Luna 16-type samplers.

Luna-1970A Launched February ?, 1970, by D-1-e from Tyuratam but failed to attain Earth orbit? Tentative identification only, but possibly attempt at Luna 16-type sampler mission.

Venera 7

Launched: 0538 GMT August 17, 1970
Vehicle: A-2-e
Site: Tyuratam
Spacecraft mass: 1,180kg at launch (capsule "about 500kg")
Destination: Venus
Mission: Lander capsule
Arrival: December 15, 1970 (landed 5°S/351° longitude)
Payload: Bus:
      Solar wind detector
      Cosmic ray detector
      Others?
    Capsule:
      Temperature and pressure? sensors
      Others?
End of mission: December 15, 1970
Notes: First Venus capsule to reach surface still transmitting

The 1970 Venus window provided the first opportunity to fly a greatly modified atmospheric capsule in an attempt to reach the surface in working order. Precise surface conditions were unknown and Soviet designers erred on the side of caution by building the capsule to withstand 540°C and 180 Earth atmospheres for 90min, adding shock absorbers to take the force of landing.

The capsule was similar in size to those of Veneras 4–6, but a smaller pressure vessel with fewer openings was carried in the egg-shaped outer layer of thermal and mechanical protection. Heat rather than pressure was expected to be the limiting factor, and provision was made to chill the interior down to −8°C before separation from the main bus. The main vehicle was very similar to those of Veneras 4–6, though only the simple sensors on the capsule have been acknowledged as the scientific payload, probably because of the telemetry system failure (see below). The main 'chute was initially reefed by nylon cord around the lines to allow the capsule to reach the lower atmosphere rapidly; it melted through as the temperature rose and permitted full deployment.

Venera 7 was accelerated out of its 182 × 202km, 51.75° Earth parking orbit by a 244sec burn of the launcher's final stage at 0659. Two course corrections were made for the first time in a Venera mission: on October 2 and November 17, at 17 and 31 million km from Earth respectively. Power was diverted to the descent capsule's batteries on December 12 and the interior was cooled in preparation for the night-side entry at 0459 on December 15.

It appears that the whole craft plunged in at 11.5km/sec and 60–70° to the horizontal. Ejection was triggered when the bus' Earth lock was broken as it was disturbed by the thickening atmosphere. Previous capsules had separated some thousands of km out but possibly the Soviet engineers wanted to leave the craft attached to its cooling system for as long as

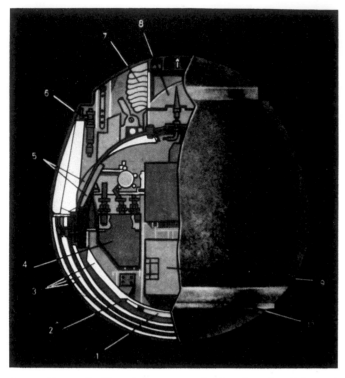

Venera 7 capsule. **Key: 1** damper mechanism **2** casing **3** thermal insulation **4** commutator unit **5** heat-exchanger **6** release mechanism for cap **7** parachute **8** antenna **9** transmitter **10** aerodynamic damper.

possible. It could however have resulted from a failure of the primary separation system, which used Earth command or an onboard timer. The holding straps were also designed to burn through as a last resort.

The dense air slowed the module to 720km/hr, and at a height of 60km the parachute system came into play. But 35min later, at 053410, all transmissions ceased. Tass announced on January 26, 1971, that a further 23min of data at 1% of normal signal strength had been recovered from what at first appeared to be pure noise. Venera 7 had indeed survived to the surface, where it found a temperature of $475 \pm 20°C$. The pressure was deduced to be 90 ± 15 atmospheres from rate-of-descent measurements derived from Doppler shifting of the transmitter frequency. The signal loss was possibly caused by the capsule falling over in windy conditions, but whatever the reason it showed that future craft could survive with less protection. The Soviets subsequently admitted that the telemetry system had become "stuck" on the temperature sensor and no other data were returned. However, the Doppler measurements were sensitive enough to register the full unfurling of the 'chute and the landing 26min later.

Kosmos 359 Mass 1,180kg, launched August 22, 1970, by A-2-e from Tyuratam into Earth parking orbit of $207 \times 298km$, 51.8°. Escape stage reignited for only 25sec, stranding Venera 7's twin in a $208 \times 890km$, 51.1° path. Re-entered on November 6.

Right: Luna 16 ground tests, with arm and drill in upright position. The descent units are attached.

Luna 16

Launched: 1326 GMT September 12, 1970
Vehicle: Proton (D-1-e)
Site: Tyuratam
Spacecraft mass: 5,727kg at launch, 1,880kg on Moon
Destination: Moon
Mission: Sample return
Arrival: September 17, 1970 (landed 0518 GMT September 20 at 0.68°N/56.30°E)
Payload: Stereo imaging system
Sampler arm with drill
Earth-return vehicle
Radiation detector(s)
End of mission: September 24, 1970
Notes: Third Soviet lunar landing, first unmanned sample return

The setbacks to and eventual cancellation of their man-on-the-Moon programme forced the Soviets to use only unmanned craft to investigate the lunar surface and acquire samples. The second-generation Lunas, launched by the A-2-e, were too limited for comprehensive studies. The much larger craft of the new generation – their launch weights were not revealed until the 1980s – could use the D-1-e for sample-return, surface rover and orbiting missions. Luna 16's landed mass was given as 1,880kg, compared with the 112kg of the second-generation Luna 13.

Building a surface rover was an outstanding achievement (NASA studied a similar "Prospector" concept), but by any standards the sample-return mission was the most difficult of all. A lander had to acquire a core sample – a surface scooping was of little scientific value – store it inside a sealed

capsule and launch it from the lunar surface at 2.7km/sec with sufficient accuracy to encounter the Earth a few days later.

Launcher failures possibly accounted for five lost opportunities, and Luna 15, which entered lunar orbit before Apollo 11, is believed to have been configured for such a mission. The success of Luna 16 14 months later allowed the Soviets to claim that unmanned lunar exploration was not only cheaper and safer than its manned counterpart but capable of producing results every bit as good.

A 185 × 241km, 51.5° Earth parking orbit was followed 70min later by injection into a translunar trajectory by the Proton's final stage (called "Block D" by the Soviets). The next day a 6.4sec burn by the main engine provided a course correction. The second-generation Lunas had corrected their paths by firing at 90° to their direction of travel to shift the arrival point, whereas the new vehicles used all three dimensions to adjust the speed of arrival as well.

Luna 16 swung around the Moon on September 17 and the main engine flared to life again to establish a circular 110km, 70°, 119min path. The propellants, hydrazine (UDMH) and nitric acid, were drawn from two pairs of identical cylindrical tanks (estimated at 1.3m long, 60cm in diameter) mounted on the craft's long sides. These detachable units also provided mountings for antennae, and one carried a nitrogen attitude-control system and the other the attitude sensors for spacecraft control all the way to descent ignition. They were attached to cylindrical sections on the descent stage, at the ends of which were four 88cm spherical tanks containing the same propellants as the insertion tanks. The short, compressible landing legs were attached to these carrier tanks to provide a maximum base of about 4m. The whole craft was about 4m high. The descent stage also had an attitude-control system, with some thrusters mounted on arms, and a radar altimeter for the landing sequence.

The eight tanks fed a single large engine (KTDU-417) in the centre of the craft yielding a variable thrust of 7.35–18.92kN for orbit insertion and descent. Though this engine could have been used for the entire sequence, two smaller chambers on either side were employed close to the surface to avoid disturbing the sampling site. Four verniers around the base periphery provided stabilisation. This basic insertion and descent arrangement was used on all of the subsequent Lunas, including the orbiters.

The craft was controlled from a pressurised toroidal section that also acted as the support and launch platform for the ascent stage. In this thermally protected enclosure were housed communications, data-processing and command electronics in addition to gyros for attitude reference and batteries (limiting these Lunas to surface lives of a few days).

Using an intermediate lunar orbit was safer than a direct descent to the surface, as the simpler second generation had done. On September 18 and 19 two burns produced paths of 15 × 110km, 70° and then 15 × 106km, 71° in preparation for the landing approach from perigee. The manned Apollos followed a complex shallow approach pattern, but all that was necessary for these unmanned vehicles was a single burn to cancel orbital velocity, followed by a vertical drop and a final burn for a soft landing. Attitude information from the insertion module system was fed into the internal gyro and the two side units were cast off just before the main engine fired at 0812 on September 20. The duration has not been released, but Luna 20 fired for 267sec to negate its 1.7km/sec orbital speed.

Luna 16 went into free fall and followed a pre-set programme modified by radar altimeter information on height and

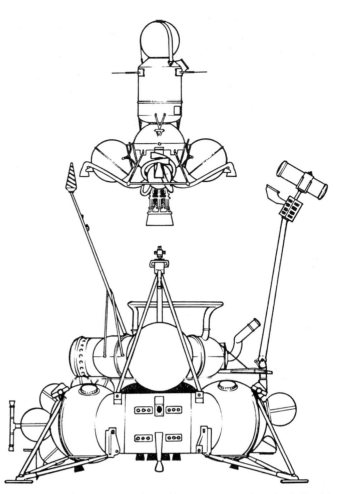

Luna 16 in surface configuration, with ascent stage detached. (David Woods)

rate of descent. Some 600m above the surface, with the craft dropping at about 700km/hr, the onboard computer ordered the engine to fire again. The radar signalled cut-off at a height of 20m and the two smaller engines ignited to complete the descent to the NE portion of the Sea of Fertility about 100km east of Webb crater. Contact speed was 9km/hr.

The first true Soviet soft-lander had come to rest at a mare site (the easiest target) in darkness. Two cameras of the Luna 9/13 type were mounted on the instrument section to swivel down and return a facsimile stereo image of the area between the two landing pads below, allowing controllers on Earth to pick the best spot for sampling. However, since this was a night-time landing and there is no indication of floodlights being carried, the cameras probably could not be used. The 90cm drill appendage was released from its stored position, latched on to the ascent stage like an upright arm, and rotated down to the target area. Electric motors could sweep it over a 100° arc and the head itself could swivel in elevation. But its fixed length limited the choice of sample sites to much less than that of the US Viking Mars landers, with their extensible arms.

A hollow rotary/percussion bit was driven 35cm into the surface over 7min and then lifted to the open hatch of the return capsule; a significant amount of soil seems to have dropped out during this movement. The drill had struck rock at the end

and controllers did not press on for fear of damaging the device. The bit and its precious cargo of 101g of lunar material were sealed aboard and Lunar 16 waited for the appropriate time to send the ascent stage on its way. All of the sampling Lunas landed around longitude 60°E, whence their ascent stages could be launched vertically into a direct path to Earth; there was no need for correction of a horizontal component at launch, nor for a course correction, thus simplifying the design.

At 0743 GMT on September 21 the 520kg ascent stage ignited its 18.8kN KRD-61 nitric acid/unsymmetrical dimethyl hydrazine main engines and four outboard verniers to pull away from its mounting and accelerate to the 2.708km/sec needed for reaching Earth. Roll jets provided spin control. A pressurised cylinder above the central tank contained the control, communications and power equipment, including the gyros and accelerometers that provided the information for cutting off the motors. The Earth trajectory achieved, the roll jets possibly induced a slow spin for temperature regulation. Telemetry was returned on 183.6MHz.

The 50cm, 39kg re-entry capsule remained attached to its host for most of the homeward journey. Towards the top it carried parachutes and descent antennae, in the mid-section was the sample and at the bottom were the batteries and transmitting equipment, an arrangement producing a displaced centre of gravity. This ensured that the greatest thickness of ablative heatshield, on the side opposite the parachute compartment, was presented to the atmosphere for re-entry.

Far left: *Luna 16 ascent-stage departure. The sample hatch is visible at top left.*

Left: *Luna 16 capsule recovery.*

Centre left: *Luna 16 return capsule showing charred ablative protection. The sample hatch appears to be at top right.*

Bottom left: *Part of the first lunar sample returned by an unmanned craft. Note the glass sphere at centre.*

Lunokhod 1.

The straps holding the capsule onto the ascent stage were severed at 0150 on September 24 and a spring probably pushed the sphere away some 48,000km out from Earth. The rest of the craft would burn up in the atmosphere. At 0510 the capsule hit the atmosphere at 30° to the vertical and travelling at 11km/sec. The heatshield glowed as temperatures reached 10,000°C and the tiny body was rapidly decelerated at up to 350g. A barometer probably provided the signal commanding ejection of the top of the sphere at a height of 14.5km, allowing the drogue parachute to spring out. The main 'chute and four beacon antennae deployed 11km up, and the surviving section of Luna 16 was spotted by aircraft as it fell towards an 0526 landing 80km SE of Dzhezkazgan in Kazakhstan.

The sample container was removed and flown to Moscow, where it was unsealed in a sterile chamber filled with inert helium. Analysis showed the composition of the dark powdery basalt material to be very similar to that returned by Apollo 12 from another mare site. It had slightly different titanium and zirconium oxide levels compared with Apollo 11's samples. Towards the bottom of the core was large-grained material, from the point where the drill had apparently come up against rock. The fine-grained material at the top had no particles greater than 3mm.

Three grammes were exchanged in June 1971 for 3g of Apollo 11 and 3g of Apollo 12 returns, half of this sample being distributed the following month for analysis by 24 US scientists.

Zond 8   Launched October 20, 1970, by D-1-e from Tyuratam as test of manned lunar craft. Flew around Moon October 24 at 1,120km and completed ballistic re-entry to splash down in Indian Ocean on October 27, returning high-quality photographs. End of Zond programme (see Zonds 4–8) following success of Apollo.

Luna 17

Launched: 1444 GMT November 10, 1970
Vehicle: Proton (D-1-e)
Site: Tyuratam
Spacecraft mass: 5,700kg? at launch (756kg Lunokhod 1)
Destination: Moon
Mission: Rover
Arrival: November 15, 1970 (landed 0347 November 17 at 38.28°N/35°W)
Payload: Lunokhod with:
        2 TV and 4 facsimile cameras
        Rifma X-ray spectrometer
        X-ray telescope (2–10Å)
        Odometer/speedometer

        Penetrometer
        Radiation detectors (cosmic rays, solar wind)
        Laser reflector
End of mission: October 4, 1971
Notes: First surface rover, fourth Soviet lunar lander

The sample-return mission of Luna 16 was restricted to investigation of a single location. Its successor replaced the Earth-return vehicle with a mobile laboratory that could be directed around the surface under control from Earth, making simple soil analyses at will. Rovers on other planets and moons in the future will need a high degree of artificial intelligence because human operators will be too distant for useful intervention, whereas the 3sec communications delay for lunar vehicles permitted real-time steering.

A 192 × 237km, 51.6° Earth parking orbit was followed by injection into a trans-lunar trajectory by the final Proton stage. Course corrections were made on November 12 and 14 before the main engine fired on November 15 to establish a circular 85km, 141°, 116min orbit. The perigee was adjusted the next day to 19km, followed by descent initiation at 0341 GMT on November 17 and a landing in a subdued ancient crater in the Sea of Rains. The descent stage and landing sequence were both very similar to Luna 16's.

Two sets of ramps were lowered – a rear set was included in case the forward exit was blocked by a boulder – and the five-man steering team working through the Yevpatoria deep-space station in the Crimea edged the rover gingerly down and on to the surface at 0628. They were in contact for only about 6hr a day, when the Moon was above the horizon.

Lunokhod 1 was essentially a pressurised magnesium alloy container on wheels, 1.35m high and 2.15m across the top of the compartment. The electronics of the communications and command system were protected in the 1 Earth atmosphere interior, with water evaporation in a heat-exchanger maintaining the temperature within 0–40°C and preferably 15–20°C. Heat was radiated away by the top, which was uncovered by a

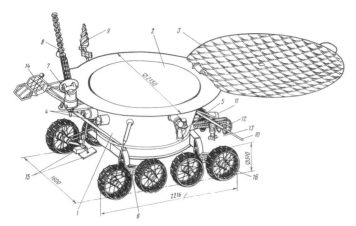

Lunokhod 1. Dimensions are in millimetres. **Key: 1** *pressurised compartment* **2** *radiator* **3** *solar panel on lid underside* **4** *cameras* **5** *rear and side-viewing cameras* **6** *wheel pair and suspension* **7** *antenna-pointing mechanism* **8** *narrow-beam antenna* **9** *low-gain antenna* **10** *rod antenna* **11** *isotopic heater* **12** *odometer* **13** *penetrometer* **14** *French laser reflectors with cover open.*

large convex lid during the day; the inside of the cover carried solar cells to charge batteries inside the compartment (this was the first Soviet lunar probe since Luna 3 to employ solar cells). A radioactive polonium-210 source outside at the back kept Lunokhod 1 warm during the cold lunar nights.

Internal gyroscopes provided information on movement to keep track of Lunokhod 1's progress, while other sensors cut off power if dangerous slopes were being attempted. Carried on the sides were four TV cameras of the Luna 9/13 type capable of returning 6,000-line images for assembly into detailed panoramas for planning purposes when Lunokhod 1 was stationary. They could scan 360° in the vertical and 180° in the horizontal to provide side, down and rear views. Two TV cameras (vidicons?) mounted on the front – central and to the right – provided stereo pictures with a 50° field of view every 20sec using 1/25th-sec exposures to prevent image smearing during travel. A steerable high-gain antenna and fixed conical low-gain antenna provided communications links.

The eight 51cm-diameter wheels, with independent suspension and each driven by an electric motor in the hub, provided a wheelbase of 1.6m and were designed to grip the dusty lunar surface with titanium blades on a fine wire mesh. Two forward and one reverse speeds were possible; these have never been quantified, though Lunokhod 1 probably moved at only a few km/hr. Changes in direction were achieved by driving the wheels on each side at different speeds or even in reverse, much like a tank. Each wheel also carried a rev counter, but slippage made this method of calculating distance travelled somewhat unreliable and a separate odometer was mounted at the back; this also acted as a speedometer.

Beside the odometer was a penetrometer to test the soil's mechanical characteristics, a cone with a cruciform being pushed into the ground and then twisted. This was used more than 500 times on Lunokhod 1. The Rifma X-ray fluorescence spectrometer mounted at the front near the surface irradiated the soil and recorded the induced radiation to identify elemental quantities of, for example, iron, titanium and aluminium, much as some of the US Surveyors did at more limited locations. Only 25 analyses were performed, since they required lengthy halts to gather data.

The cover on a 3.7kg French experiment above the forward cameras was opened when Lunokhod 1 was parked during the long nights. This contained 14 10cm silica glass prisms designed to bounce back 10^{-8}sec pulses of ruby laser light fired from the Crimean (Soviet) and Pic du Midi (French) observatories. This was first achieved on December 5 and 6. Measuring the time of travel allowed the Earth–Moon distance to be calculated to within 30cm (Apollo astronauts also left laser reflectors on the surface).

Lunokhod 1 had a design life of three lunar days but in the end it operated for 11. The following is a brief chronology of its lengthy career:

Day 1 November 17–22: 197m travelled as controllers carefully tested their skills and the vehicle in low gear. **Day 2** December 10–22: 1,522m travelled heading south. Sensors at one point cut power on a 27° slope. Crossed several craters. Parked 1.4km from lander. **Day 3** January 8–20, 1971: 1,936m travelled. Parked January 13–15 when high Sun angles washed out surface features. Headed north and halted back at lander. **Day 4** February 8–19: 1,573m travelled. Moved north, finding 20m young craters. Observed 3hr solar eclipse February 10, temperature dropped to −100°C from +138°C in a few hours. Parked on a crater rim 1km north of lander. **Day 5** March 9–20: 2,004m travelled. Moved NW and investigated large crater, making two X-ray soil analyses. **Day 6** April 8–20: 1,029m travelled. Traversed boulder field with 3m objects, wheels sank in 20cm of dust, 30° slopes negotiated. **Day 7** May 7–20: 197m travelled. Performed mainly static experiments, possibly because of equipment problems. **Day 8** June 5–18: 1,559m travelled in NE direction. After this session the vehicle seemed to deteriorate. **Day 9** July 4–17: 220m travelled northwards. Crossed 200m crater. **Day 10** August 3–16: 215m travelled. **Day 11** September 1–15: 88m travelled.

The end-of-life announcement was made on October 4, 1971, the anniversary of Sputnik 1, but failure probably occurred earlier, partly because the nuclear heater had decayed and allowed the internal temperature to drop dangerously low during the night. Controllers had the foresight to park Lunokhod 1 so that the laser reflectors were left in a usable position. The 322 Earth days of operations covered 10,540m and returned more than 20,000 TV images and 206 detailed panoramas.

See Luna 16, for descent-stage description, and Luna 21.

Mariner 8   Launched 0111 GMT May 9, 1971, by Atlas-Centaur 24 from ETR 36A to map Mars from orbit. The spacecraft fell into the Atlantic 1,450km SE of the Cape after a circuit chip failed in the Centaur autopilot and left the engines without pitch control.

Kosmos 419   Mass 4,650kg, launched May 10, 1971, by D-1-e from Tyuratam but failed to leave Earth orbit. Intended as Mars orbiter and landing capsule (see Mars 2/3).

Lunokhod's own view of its landing stage with deployed ramps.

Mars 2

Launched: 162249 GMT May 19, 1971
Vehicle: Proton (D-1-e)
Site: Tyuratam
Spacecraft mass: 4,650kg at launch (450kg landing capsule)
Destination: Mars
Mission: Orbiter/lander
Arrival: November 27, 1971 (lander crashed 45°S/58°E)
Payload: Orbiter:
> Film-TV system
> Infra-red radiometer (8–40μm: surface temperature and atmospheric water vapour)
> 3.4cm radio receiver (density and temperature of subsoil, 3–50cm deep)
> Ultra-violet photometer (atmospheric emissions from atomic hydrogen, oxygen and argon)
> Visible light photometer
> Magnetometer
>
> Lander:
>> Mass spectrometer
>> Temperature and pressure sensors
>> Anemometer
>> TV cameras
>> Soil mechanical/chemical analyser

End of mission: March 1972 (announced formally August 24, 1972)
Notes: First object on Mars, second Mars orbiter (first USSR)

The favourable Mars window of 1971 allowed the Soviets to fly the full three-axis-stabilised orbiter/lander configuration of their second-generation probes. The single 1969 attempt was probably a flyby/lander, since the window did not allow sufficient propellant for establishment of an orbiter.

Mars 2 was injected into a Type I transfer orbit at 1759 GMT on May 19, the Soviets announcing that it would conduct research "about Mars" and collect solar wind and cosmic ray data en route. This was the first of the new Mars craft to break

Above Mars 2/3 descent section minus the heatshield.

Below: Cutaway of the Mars 2/3 capsule. The Soviet caption to this drawing makes it clear that Mars 6 was of identical design. **Key: 1** control-system altimeters **2** shock-absorber **3** telemetry equipment **4** radio-control system **5** antennae **6** antenna feeders **7** radio **8** scientific equipment **9** cameras **10** locking device for petals **11/12** protective shell **13** thermal insulation **14** heat-insulating upper body (when open on surface) **15** pyrotechnic devices for opening petals **16** petals opened after landing **17** pressurised bag (part of opening system?) **18** protective casing **19** shock-absorber **20** charge for jettisoning protective cover **21** automatic control system **22** batteries **23** "air pressure receiver" (for pressure measurements during descent?).

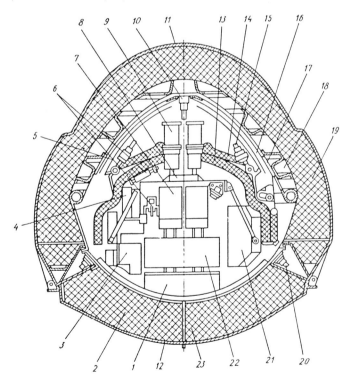

Left: Mars 2/3, with landing capsule housed under the conical heatshield. The parachute torus is visible.

65

free of Earth parking orbit; the subsequent Mars 3–7 and Venera 9–18 missions used the same basic model. It was four times heavier than the previous Veneras and equipped with new navigation, control, communications and propulsion systems. With its 2.3 × 1.4m solar panels deployed, Mars 2 spanned 5.9m; total height was 4.1m. A 1.8m-diameter cylinder formed the main body and housed the propellant tanks, flaring to a base diameter of 2.3m around the main motor. The 9.86–18.89kN KTDU-425A engine, drawing on nitric acid and amine-based propellants, was also used in the later generation Veneras. Most of the electronics and instruments were mounted on or carried inside this pressurised conical section. Dominating one side was a 2.5m-diameter high-gain antenna; a conical command/relay antenna was carried on the reverse side of each solar panel. Inboard of the panels were radiators to dissipate waste heat. Capping the structure was a 1.2m-diameter landing capsule nestling under a 2.9m-diameter conical heatshield.

Mars 2 made its first trajectory correction at 0130 GMT on June 17 while some 7 million km from Earth, and a second on November 20 just before Mars arrival. Unlike the later US Vikings, Mars 2 had to release its lander before entering orbit and could not wait in orbit for a raging planet-wide dust storm to clear. The 450kg sterilised capsule in its 185kg protective shell separated from the bus 4½hr out from Mars and fired a

Soviet representation of Mars 3 after touchdown.

small solid-propellant motor to shift its path to the aim point. The main craft made a manoeuvre to ensure that it missed the planet.

The automatic sequence called for the capsule to hit the atmosphere at 6km/sec and mortar-eject a drogue 'chute to extract the main canopy, packed in a torus girdling the sphere, while it was still travelling supersonically. The heatshield should then have fallen away to allow a radar altimeter to ignite small solid retromotors above the capsule for a soft landing three minutes after atmospheric contact. Another small rocket would have pulled the detached and collapsing parachute sideways to prevent it from covering the lander. The capsules of this Mars series were very similar to those of Lunas 9/13, with four petals opening to establish an upright position on the surface. The small omnidirectional antenna would then have beamed stereo image data from the twin cameras up to the orbiter for relay to Earth. Soviet statements indicated that there were devices for studying the mechanical and chemical properties of the soil, suggesting the presence of a penetrometer (as carried on Luna 13?) and a gamma or X-ray detector.

Unfortunately, possibly because of the storm or some vehicle failure, the Mars 2 capsule appears to have crashed. The main craft fired its main engine at 2319 on November 27 to establish a 1,380 × 25,000km, 48.9°, 18hr orbit. It carried telephoto (4° field of view) and wide-angle (52mm focal length) cameras to image the surface using the Luna 3 method. Twelve frames of exposed film were to be developed and scanned at 1,000 lines (1,000 pixels/line) for transmission to Earth, but it appears that the dust storms severely reduced the quality. The Soviets later claimed that orbiter imaging had been assigned only a "subsidiary role". Three pictures from Mars 2/3 were shown on Moscow TV on January 22, 1972.

Their landers having failed, Mars 2 and 3 concentrated on surface and atmospheric studies. They discovered atomic hydrogen and oxygen in the upper atmosphere and found a surface temperature range of 13°C at midday to −110°C at night, although areas on the night side could be 20–25°C warmer than their surroundings. Surface pressure was 5.5–6mb and water vapour, as expected, was scarce. See Kosmos 419 and Mars 3.

Mars 2/3 descent. The lowered device could have contained a contact probe to ignite the retromotors clustered below the parachute.

Mars 3

Launched: 152630 GMT May 28, 1971
Vehicle: Proton (D-1-e)
Site: Tyuratam
Spacecraft mass: 4,650kg at launch (450kg lander)
Destination: Mars
Mission: Orbiter/lander
Arrival: December 2, 1971 (landing at 45°S/158°W at 1349 GMT)
Payload: As Mars 2 + French Stereo 1 solar radio receiver (1m wavelength)
End of mission: As Mars 2
Notes: Second object on Mars (successful landing?); third Mars orbiter (second USSR)

Mars 3 was the third of the Soviet craft to be launched during the favourable 1971 window and the second to reach the Red Planet.

A course correction was made with the main engine at 0210 GMT on June 8. At 0914 on December 2, its batteries freshly charged from the main craft, the landing section separated; 15min later it fired its solid-propellant motor to head for the target site. The automatic landing sequence appears to have achieved a 75km/hr touchdown and the craft's timer ordered a panoramic scan of the surroundings just 90sec later. However, only 20sec of TV signals, showing no contrast at all, were recorded by the orbiter for relay to Earth before transmissions were "suddenly discontinued", according to the Soviets.

The main section executed a braking burn to enter a 1,530 × 214,500km, 60°, 12-day 16hr orbit. Its imaging systems showed the landing area to be under a thick dust cloud. High winds could have silenced the small descent craft, but there have been suggestions that the lander was completely successful and that the problem lay in the orbiting relay. See Mars 2 for vehicle description.

Mariner 9

Launched: 2223 GMT May 30, 1971
Vehicle: Atlas-Centaur 23 (Atlas No 5404C)
Site: ETR 36B
Spacecraft mass: 998kg at launch (565kg in orbit)
Destination: Mars
Mission: Orbiter
Arrival: November 14, 1971
Payload: TV system (25.8kg)
Ultra-violet spectrometer (15.9kg)
Infra-red radiometer (3.6kg)
IR interferometer spectrometer (23kg)
End of mission: October 27, 1972
Notes: First artificial satellite of Mars (first planetary orbiter)

Before the Viking landers could be flown to search for life on Mars, it was necessary to map the planet from orbit. NASA announced on November 14, 1968, that JPL was to conduct a $120 million project to launch two orbiters in spring 1971 to take advantage of the most favourable launch-window conditions in the 15-year cycle. Whereas Mariners 4, 6 and 7 had imaged only a tiny proportion of the Martian surface, Mariner 8

would enter a 12hr, 80° orbit with a perigee of 1,200km to spend three months mapping 70% of the surface. Towards the end of the 28-day window Mariner 9 would enter a 20.5hr, 50° path, approaching the planet to within 885km. It would be concerned with transient features, studying the atmosphere, polar caps, clouds and haze layers.

The identical spacecraft would each carry an IR radiometer to map the temperature of the surface, partly in search of warm spots that might favour life. The other UV and IR instruments would study the composition and vertical profiles of the atmosphere, particularly water vapour. But the main equipment on the scan platform was the twin TV camera. The 508mm-focal-length, f/2.35 narrow-angle camera could produce a resolution of up to 1km; the 50mm, f/4 camera with its filter wheel for colour and polarisation studies covered wider areas. Each image took 42sec to generate, the 700 lines off each vidicon broken up into 832 9-bit pixels per line and stored on 175m of tape at 132kbits/sec. Subsequent playback

Below: Mariner Mars 1971.

Bottom: Underside view of Mariner 8/9.

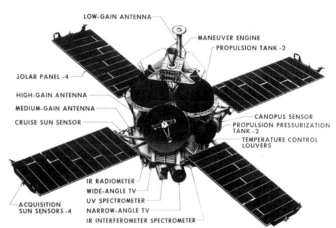

NOTE: PROPULSION MODULE AND SCAN PLATFORM INSULATION BLANKETS NOT SHOWN

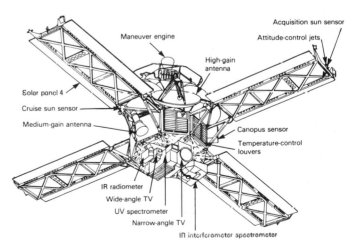

Note: Propulsion module and scan platform insulation blankets not shown.

to Earth would take place at up to 16.2kbits/sec via the 20W, 100cm-diameter aluminium-honeycomb high-gain S-band antenna.

The basic spacecraft closely resembled the Mariner 6/7 vehicle, with large propellant tanks added on top to permit the 1,600m/sec retroburn needed for Mars orbit. The single 1,340N engine, capable of six 0.4sec–15min burns, was fed from two 76cm spherical titanium tanks holding a total of 463kg of hypergolic monomethyl hydrazine and nitrogen tetroxide propellants, and pressurised by nitrogen from two tanks. It turned out that the main engine performed·slightly better than expected, and the onboard accelerometer had to cut if off early during the long braking burn of November 14. Twelve nitrogen thrusters mounted on the ends of the solar panels provided attitude control, along with gyros and Sun and Canopus sensors.

The octagonal 18kg magnesium framework of the main body contained the electronics, with the outer faces providing thermal control; six of the faces carried thermal louvres designed to open at an internal temperature of 29°C. Thermal blankets covered the main propulsion system and instrument section. The 10.4kg Central Computer and Sequencer, with its magnetic-core memory of 512 22-bit words, could be updated in flight to provide the greater flexibility demanded by extended orbital operations. A sequencer acted as a backup to issue commands at fixed times if the primary system failed.

Power was produced by 17,472 solar cells mounted on four 0.9 × 2.14m panels deployed after launch to generate 800W at Earth and 450–500W at the distance of Mars. A single rechargeable nickel-cadmium battery provided 600W-hr during manoeuvres that put the panels out of solar alignment.

Fully deployed after launch, each Mariner was 2.3m tall and spanned 6.9m.

The launch failure of Mariner H (8) left its twin facing a combined mission around Mars. It would now aim for 65° inclination in order to map much of the surface and make repeated visits to some sites every 17 days to study transient phenomena. The modified Centaur (see Mariner 8) this time burned its two main engines for 454sec in a direct ascent to start Mariner 9 on its Type I journey at 11.05km/sec. On June 4, 1971, 1.35 million km from Earth, the main engine was fired for 5.11sec to produce a ΔV of 24.3km/hr and a Mars miss distance of 1,600km. This first correction was so accurate that a second, on October 26, was cancelled. The motor next roared into life at 0018 GMT on November 14 with the spacecraft 2,753km from Mars, burning for 915.6sec and consuming 431kg of propellant to brake Mariner into a 1,398 × 17,916km, 64.5°, 12hr 34min orbit. The path was trimmed to 11hr 57min, 1,394 × 17,144km, ·64.34° on November 15 with a 6sec, ~15m/sec burn.

The first images had already been returned on November 10 from a distance of 860,000km, but they were disappointing. Earth-based astronomers witnessed a huge, Mars-girdling dust storm rise in late September, and all the cameras could distinguish for weeks were four dark spots in the Tharsis region and the bright south polar cap. Tracking quickly revealed a 27km equatorial bulge, only 18km of which could be accounted for by the planet's rotation. Was there an active interior? For the first time, a spacecraft detected water vapour in the atmosphere, over the south pole.

On November 26/27 man's first close-up images of Deimos, and then Phobos on November 29/30, showed the two tiny moons to be irregular and covered in impact craters. They were certainly not the artificial satellites of an advanced civil-

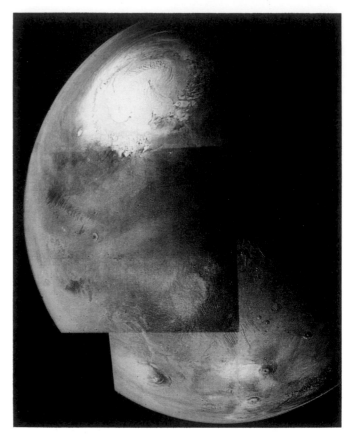

Above: *The northern Martian hemisphere imaged by Mariner 9 on August 7, 1972. Olympus Mons is at lower left and the western end of Valles Marineris is just visible at bottom right. The north polar cap is shrinking as summer approaches.*

Below: *Mariner 9's arrival in orbit was greeted by a planet-wide duststorm, but by late January 1972 the atmosphere had cleared sufficiently for this mosaic of the huge Olympus Mons to be produced. The central caldera is approximately 65km across.*

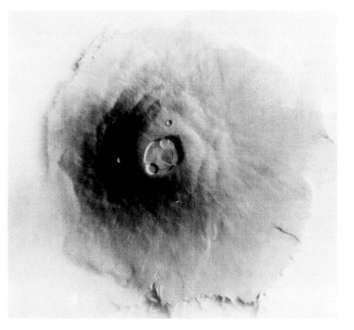

isation, as had been claimed in some quarters. Instead, they appeared more like captured asteroids.

The dust storm, sometimes reaching 70km high, gradually subsided and on January 2, 1972, the mapping sequences began in earnest. Geologists were astonished at the diversity of the features that lay before the two cameras, the three previous Mariners having all flown over particularly Moon-like areas. The four spots in the Tharsis-Amazonis region turned out to be huge volcanoes. The largest of them, now officially named Olympus Mons, dwarfs anything the Earth has to offer. Some 25km high, it has a base extending to a diameter of 600km, with lava flows everywhere. The three others sit in a line 1,000km to the SE on the newly discovered bulge.

On January 12 the first section of the Valles Marineris rift valley was seen to the SE of the volcanoes. It proved to be 4,000km long, 100km wide and 6km deep in places, stretching over a fifth of the globe's circumference 10–15° south of the equator. Earth's Grand Canyon would qualify as a small tributory channel to this mighty cleft.

These features suggested that Mars is a geologically active planet, the Tharsis bulge and the nearby rift possibly indicating the beginning of plate tectonics. The southern half of Mars is ancient cratered terrain (at least 3.9×10^9 years old) surrounding the large Hellas (2,000km diameter) and Argyre (900km) impact basins. The north appears to be eroded or covered by dust. There are also sinuous channels that exhibit characteristics of being cut by free-flowing water in the recent past: does Mars undergo periods of rainfall? The atmosphere of carbon dioxide is clearly too thin to support free surface water at present. Mariner's radio occultations showed pressure to be a mean of 2.8mb at the equator, rising to a high of 8.9mb at mid-latitudes. The south polar cap was interpreted as a 300km-wide water-ice layer in summer, hidden by a 3,000km-diameter covering of frozen carbon dioxide in winter.

Mariner 9's observations transformed our view of the Red Planet. It was now regarded as more Earth-like, with a greater chance of harbouring life. NASA adjudged the mission to have achieved its primary objectives by February 11, 1972.

Mars was in conjunction with the Sun from August to October 12, 1972, and ground controllers were unable to command Mariner 9. By then, however, most of the surface had been mapped at 1–2km resolution and 2% of it at 100–300m. The northern regions were completed after the planet emerged from conjunction, but on October 27, during the 698th revolution and after 349 Earth days of orbital operations, the nitrogen tanks ran dry and Mariner began to tumble. Fifteen pictures were left untransmitted on tape but 7,329 had been received, and the last of 45,960 commands turned off the transmitter for good. The orbit was believed to be stable for 50 years, so satisfying international planetary quarantine requirements. Apart from their direct scientific value, the images could now be used to select suitable landing sites for the two Vikings of 1975–6, and five preliminary candidates were named on November 6, 1972.

Apollo 15 Particles and Fields Subsatellite

Launched: July 26, 1971 (ejected from Apollo 2013 GMT August 4)
Vehicle: Saturn V (No 510)/Apollo CSM-112
Site: ETR 39A

The Apollo 15 SIM bay with the closed subsat container (arrowed).

Spacecraft mass: 36kg
Destination: Moon
Mission: Orbiter
Arrival: July 29, 1971
Payload: Magnetometer
Particle telescopes
End of mission: February 3, 1972
Notes: First man-deployed planetary orbiter

Astronauts Scott, Irwin and Worden ejected the 78cm-long, 36cm-diameter hexagonal subsatellite into an initial orbit of 102 × 139km, 28.5° shortly before they left lunar orbit to return home following the fourth manned landing. A spring pushed the simple, TRW-built satellite out of the Service Module experiment bay, imparting a 12rpm spin for stabilisation once the three end-mounted 1.83m booms had deployed. Each of the six faces carried solar cells to produce a total of 24W; 11 silver-cadmium batteries were also aboard. There was no active attitude control.

The satellite housed two magnetometers on one boom (the others holding only balancing masses) and sets of particle detectors to map the Moon's remnant magnetic field and its effect on particles in the solar wind and the Earth's magnetic tail. It was designed to work at altitudes between those covered by the Apollo surface experiments and the high Explorer 35 (qv). Since the 2hr orbit took the satellite over the far side there was a 49,152-bit data store, dumped at 512 bits/sec to ground stations.

The magnetic field mapping was performed best during the five days a month that the Moon spends in the Earth's quiet magnetic tail. It showed that the remnant field is quite irregular, especially on the far side, and can be correlated roughly with topographical features. The largest variation, 10^{-5} gauss, was found at 180° longitude near the crater Van der Graaf. Particle energies and directions of travel for electrons and protons were measured in the 20–700keV range with two solid-state telescopes and, for electrons only, in the 0.53–15keV range by five detectors. They showed that, whereas the Earth's magnetic field deflects the incoming solar wind into a tail, the Moon merely acts as a physical barrier

The Apollo 15 subsatellite drifts away from Endeavour *following release.*

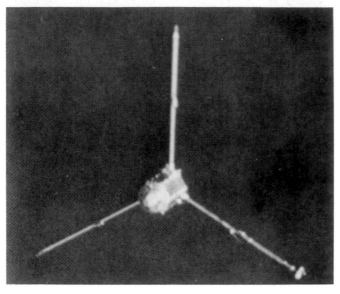

because of its weak field, creating a "hole" in the wind behind it. This was discovered by Explorer 35. The Apollo subsatellite showed that the void extends down near the surface but its edge is not simply defined.

Accurate tracking of the two Apollo satellites and Lunar Orbiter 5 produced maps of lunar mass concentrations (mascons), showing that they are caused by lava flooding of basins rather than buried material, as previously suspected. The far side, with its unfilled basins, caused only weak variations in the orbits.

A apparent electronics failure in the telemetry system on February 3, 1972, caused the loss of major portions of data, effectively ending the mission. See Apollo 16 subsatellite.

Luna 18   Sample-return mission launched 1341 GMT September 2, 1971, by D-1-e from Tyuratam into 186 × 242km, 51.6° Earth parking orbit. Course corrections were executed on September 4 and 6 before insertion burn at ~2100 GMT over lunar far side established 100km circular, 35°, 119min orbit. Descent initiated on September 11 after 54 revs, but contact lost at 0748 just before expected landing at 3.57°N/56.5°E near edge of Sea of Fertility some 120km north of Luna 16 site. Soviets explained crash as due to rugged nature of site. See Luna 20.

Luna 19

Launched: 1000 GMT September 28, 1971
Vehicle: Proton (D-1-e)
Site: Tyuratam
Spacecraft mass: 5,700kg? at launch (1,900kg? in lunar orbit)
Destination: Moon
Mission: Orbiter
Arrival: October 3, 1971
Payload: Imaging system
Meteoroid detectors
Cosmic ray detectors
Magnetometer
Radiation detectors
Radar altimeter
Gamma-ray spectrometer
End of mission: October 1971
Notes: First Soviet third-generation orbiter

The third generation of Soviet Luna vehicles initially concentrated on sample-return missions, a single successful surface rover interrupting the series. Luna 19 was the first designed as a dedicated orbiter and represented a major step forward from the limited Luna 12/14 orbiters of the previous generation. The US Explorer and Lunar Orbiters had relatively simple objectives, partly because the Service Modules of Apollos 15–17 carried an extensive range of instruments for probing the Moon and its environs. The Soviets did not enjoy that manned option and Lunas 19/22 became their main platforms.

The heavy orbiters were based on a wheel-less Lunokhod acting as the pressurised instrument carrier atop standard orbit-insertion and landing units (see Luna 16). These insertion tanks would not be jettisoned, as on landing missions, because their attitude sensors and gas jets were essential throughout the orbiting mission. The insertion tanks were now the main propulsion unit (for the ~900m/sec braking into lunar orbit) since the ~2km/sec capability of the descent tanks was not required. Luna 19 made few orbital changes and even the more numerous manoeuvres executed by Luna 22 amounted to only about 350m/sec, so the tanks were probably underloaded for these two missions. At the same time, extra nitrogen tanks were probably carried, since attitude-control propellant is a major limit on the life of a probe or satellite. Lifetimes of the earlier Luna orbiters were determined by the battery supply, so the solar cells of the open Lunokhod lid were used to provide power for the new craft.

An Earth parking orbit of 172 × 260km, 51.6° was followed by course corrections on September 29 and October 1 after Soviet observatories had pinpointed the vehicle in space. The braking burn on October 2 established a circular 140km, 40.58°, 123.75min orbit. A correction four days later resulted in a 135 × 127km path. One of Luna 19's main functions was surface photography, but it is not clear when imaging, using the film-scan method pioneered by Lunas 3/12 and Zond 3, began. The orbit was rather lofty for high-resolution photography, but a Tass statement of October 19 said that photographic activities were under way.

Another puzzle is a February 1972 report stating that images were still being produced. It is possible that Luna 19 concentrated on general mapping (a March 19 statement mentioned panoramic highland views around 30–60°N/20–80°E) while Luna 22 was given more specific targets to cover from its 25 × 244km orbit. The late imaging by Luna 19 could have been a one-off demonstration of the system some months after completion of the main mapping sequence; Luna 22 certainly did this after a year of operations.

Orbital tracking in order to map variations in the gravitational field was also given as a major objective. Luna 19 had completed 1,358 revolutions by January 31, 1972, and 1,810 by March 10. The orbit was adjusted at some stage to 77 × 385km, but there are no indications of other major manoeuvres even though the basic heavy Luna would have had sufficient capacity.

In a radio experiment in May/June 1972 signals with wavelengths of 8cm and 32cm were beamed to Earth as Luna 19 slipped behind the Moon's disc. The differential effect on their propagation revealed a concentration of charged particles at an altitude of 10km, believed to result from the interaction of solar radiation with the surface. The gamma-ray spectrometer would have been able to establish the elemental surface composition by recording characteristic gamma-ray emissions resulting from natural radioactivity and cosmic ray interaction with the regolith. The radar altimeter would not only have mapped topographical variations but also allowed deductions on the nature and depth of the regolith itself. Both of these instruments would have benefited from a low orbit.

Studies of the solar wind were co-ordinated with observations from Mars 2/3 circling the Red Planet and from Veneras 7/8. A surge in activity was picked up by Luna 19 in December 1971. The magnetometer showed the interplanetary magnetic field to be considerably weaker on the night side, and the orbit allowed study of lunar effects on the Earth's magnetic tail.

Tass reported on October 3, 1972, that Luna 19, having completed more than 4,000 revolutions, was nearing its end; it probably ceased operations later that month. See Lunas 16 and 22 for spacecraft description.

Luna 20

Launched: 0328 GMT February 14, 1972
Vehicle: Proton (D-1-e)
Site: Tyuratam
Spacecraft mass: 5,700kg? at launch (1,880kg? on Moon)
Destination: Moon
Mission: Sample return
Arrival: February 18, 1972 (landed 1919 at 3.53°N/56.55°E)
Payload: As Luna 16
End of mission: February 25, 1972
Notes: Fifth Soviet lunar landing; second Soviet sample return

Luna 16 made the first unmanned sample return from a mare site in September 1970, but Luna 18 crashed in its attempt at an older highland location. Luna 20 was launched into a 191 × 238km, 51.5° Earth parking orbit to begin the second assault on the same spot: it was important to acquire highland material for comparison with the Luna 16 basalt.

A course correction followed on February 15 and insertion into a circular 100km, 65°, 118min lunar orbit on February 18. Luna 20 lowered itself into a 21 × 100km path the next day. A 267sec burn by the main engine (this is the only mission for which the duration of the deorbit burn has been given) at 1913 GMT on February 21 cancelled the orbital velocity and allowed the craft to drop vertically under gravity. At a height of 760m (160m higher than for Luna 16) the radar altimeter signalled the main engine to reignite, cutting it off 20m up for the two secondaries to complete the descent.

Luna 20 had come down just 1.8km from the Luna 18 site near the crater Apollonius on the edge of the Sea of Fertility. The daylight touchdown (Sun elevation 60°, surface temperature 120°C) provided better lighting conditions than on Luna 16 for ground controllers to select the best drilling point. On command, the head came down and easily bit through the upper layer of loose rock until it hit hard material 10–15cm down. Three times in the seven-minute drilling sequence the motor cut out automatically because of overheating, and it is possible that an incomplete sample – perhaps only 50g – was obtained.

The head deposited the core in the capsule and Luna

Impression of Luna 20 ascending from rugged lunar surroundings.

marked time for almost a day, waiting for the appropriate terrestrial alignment, before the ascent stage ignited at 2258 on February 22. The small unit accelerated up to 2.7km/sec after only 27hr 39min on the surface, heading vertically for an uncorrected trajectory back to Earth. The capsule was apparently very similar to Luna 16's, with a slight internal rearrangement.

The small sphere separated on February 25 some 52,000km out from Earth. It hit the atmosphere at 60° to the vertical instead of its predecessor's 30°, reducing the severe deceleration but resulting in higher temperatures because of the lengthier atmospheric traverse. Even so, only 5mm of ablative coating was lost from the leading side and almost none at all from the rear. Aircraft in the 80 × 100km target area picked up the radio signal, but the severe weather prevented immediate recovery after the 1912 GMT landing on an island in the Karkingir River 40km NW of Dzhezkagan (48°N/67.6°E). The bright-orange parachute was clearly visible after daybreak and the precious cargo was soon on its way to Moscow for analysis.

This highland sample proved to consist largely of ancient anorthosite, whereas the mare examples of Luna 16 and Apollo contained only 1–2% of this material. The largest particle size was >1mm, the average 70–80μm density was 1.1–1.2g/cc. It was generally lighter-coloured than Luna 16's material and exhibited the largest content of aluminium oxides and calcium oxides of all the lunar samples analysed up to that time. Most exciting for the Soviet investigators was the discovery of non-rusting iron of very high purity.

Samples were subsequently exchanged with US and French scientists. Two grams were handed over to NASA in a formal ceremony in Moscow on April 13, 1972, the Americans responding with 1g of Apollo 15 soil.

See Lunas 16 and 24.

Pioneer 10

Launched: 0149 GMT March 3, 1972
Vehicle: Atlas-Centaur 27 + TE-M-364-4 (Atlas No 5007C, 400th Atlas launch)
Site: ETR 36A
Spacecraft mass: 258.0kg at launch
Destination: Jupiter, interstellar space
Mission: Flyby
Arrival at target: December 5, 1973, closest approach 130,000km at 0225 GMT
Payload: Imaging photopolarimeter (4.3kg)
Infra-red radiometer (2.0kg)
Magnetometer (2.6kg)
Plasma analyser (5.5kg)
Charged-particle composition instrument (3.3kg)
Ultra-violet photometer (0.68kg)
Cosmic ray telescope (3.2kg)
Geiger tube telescopes (1.6kg)
Jovian trapped-radiation detector (1.77kg)
Asteroid-meteoroid detector (3.3kg)
Meteoroid detector (1.68kg)
End of mission: 1994?
Notes: First Jovian flyby, first probe to leave solar system, first crossing of Asteroid Belt

The dual Pioneer outer-planet probes were designed to blaze the trail for later, more capable spacecraft such as Voyager. Whereas Mariner-type craft were three-axis-controlled, with some instruments carried on scan platforms, the Pioneer class was spin-stabilised with body-mounted equipment.

Pioneers 6–9 were launched between December 1965 and November 1968 into Earth-type orbits spread around the Sun to provide information on radiation and magnetic fields in interplanetary space. The new craft would head outwards instead to become the first vehicles to penetrate far beyond Mars' orbit. Their primary objective was to survive Jupiter's intense trapped radiation, having first endured seven month journeys through the dust hazards of the 280 million km-wide Asteroid Belt. If all went well, two Voyagers would follow several years later to conduct the surveys proper.

The Pioneer project was approved by NASA in February 1969. A 66,700N TE-M-364-4 solid-propellant motor – a modified version of the Surveyor lunar soft-lander retromotor – would be used as an Atlas-Centaur third stage for the first time, allowing a small vehicle to be launched during the February 25–March 20, 1972, window for the 997 million km journey. The near-identical Pioneer 11 would follow as back-up a year later.

Pioneer, built by TRW and managed by NASA Ames, was basically a hexagonal main compartment 35½cm deep and 71cm on a side containing the electronics and topped by a 2.74m 8W S-band antenna made of aluminium honeycomb. Since the Sun's power at Jupiter would be only 4% of that available at Earth, two pairs of SNAP-19 radioisotope thermoelectric generators (RTGs) powered by the decay of plutonium 238 were carried on the ends of 3m booms mounted 120° apart to provide 140W at encounter (the 29.9kg of instruments required 25W). A third boom, 6.55m long, carried the magnetometer. Course corrections and attitude control were achieved by six hydrazine jets with 1.8–6.2N variable thrusts in pulsed or continuous mode, with some firing through cutouts in the antenna rim, providing a total ΔV capability of 675km/hr.

Below: Pioneer 10 seen before mating with its Centaur upper stage.

Bottom: Pioneer 10.

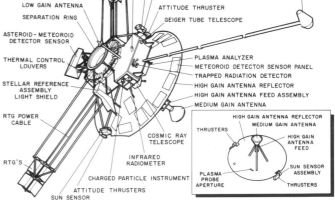

The 11 instruments were designed to return data on the magnetic fields, energetic particle radiation and dust populations in interplanetary space from well before encounter to years after solar system exit. At encounter the IR radiometer would generate thermal data and the UV photometer would help to determine the hydrogen and helium composition of the atmosphere. But the most spectacular results were expected to come from the imaging photopolarimeter (IPP). A 2½cm-diameter Maksutov telescope would scan the planet and satellites as Pioneer rotated at 4.8rpm to allow red and blue images to be build up strip by strip from the two sets of channel detectors. Addition of a green component during ground processing would yield full-colour pictures.

Pioneer 10 was launched on a Type I direct-ascent trajectory and accelerated to 51,670km/hr – the greatest speed of any spacecraft at the time – crossing the Moon's orbit after only 11hr. Launch accuracy was such that the corrective burn on March 7 amounted to only 50.4km/hr.

Pioneer 10 entered the Asteroid Belt in mid-July 1972 and emerged the following February. Dust flux rates were much lower than expected, allaying fears of damage on future missions: particles as small as 0.05cm could have been fatal in a collision at 15 times the speed of a bullet. Between launch and encounter only 55 impacts were recorded by the 234 meteoroid detector cells mounted on the dish underside. Each was pressurised with a nitrogen/argon mixture that leaked and triggered the detection system when the cell wall was breached. The 13 panels covered a total area of 0.6m². Larger material was registered by the four 20cm non-imaging telescopes and photomultipliers of the Asteroid-Meteoroid Detector; several 10–20cm objects were detected in the belt.

On November 6, 1973, 25 million km out from Jupiter, the long-range imaging tests began. The orbit of the outermost known moon, Sinope, was crossed two days later. On November 26, 109 Jovian radii away from closest approach, Pioneer 10 crossed Jupiter's bow shock, where the solar wind interacts with the planet's magnetic field, cutting particle speed from 450km/sec to 225km/sec. By December 2 the IPP was returning pictures with higher resolution than any obtained from Earth-based telescopes. But the opportunity for seeing moon Io was lost, as were some images at closest Jupiter encounter, because energetic electrons and protons trapped in the magnetic fields generated spurious signals in the craft's electronics. Nonetheless, more than 300 images were obtained, including views of Jupiter's terminator and the Great Red Spot. The spot was seen to be an atmospheric feature resembling a large group of thunderstorms rising several km above the highest clouds. A smaller red spot was discovered at about the same latitude in the northern hemisphere, but it had dispersed by the time of Pioneer 11's visit.

Tracking revealed the planet's mass to have been underestimated by a quarter-Moon mass, and analysis of the probe's S-band signals propagating through the upper atmosphere as it slipped behind the limb allowed density profiles to be determined. Io was found to possess an ionosphere and to be embedded in a cloud of hydrogen extending a third of the way around its orbit. The magnetometer showed Jupiter's dipole moment to be 19,000 times stronger than Earth's. Several of the energetic-particle detectors approached saturation: the electron flux was 10,000 times greater than in the Earth's van Allen belts. The infra-red data showed that Jupiter emits 1.7 times the radiation it receives, and revealed an atmospheric helium:hydrogen ratio of 0.14 ± 0.08, close to the Sun's value of 0.11. The planet appeared to be almost entirely fluid, possibly with a small several-Earth-mass core.

The encounter formally ended on January 2, 1974. Pioneer 10's success allowed its twin to venture three times closer before being targeted for the first Saturn flyby. Meantime, the earlier probe's investigations are continuing. Ground stations should be able to pick up data until at least 1994, by which time Pioneer 10 will be well down the Sun's magnetic tail. It so happens that the craft is leaving the solar system (it crossed the orbit of Neptune, then the outermost planet, on June 13, 1983) along the line of the heliosphere's tail and could cross over into interstellar space (at least 50 AU from the Sun) as early as April 1990.

Since the two Pioneers are departing in almost opposite directions, precise tracking could identify a new trans-Neptunian planet or even a dark solar companion by revealing differential gravitational effects on the trajectories. Once clear of the Sun's gravity, Pioneer 10 will recede from Earth at

Top: *Jupiter imaged by Pioneer 10 from a distance of 2,560,000km. Moon Io is also visible.*

Above: *On June 13, 1983, Pioneer 10 passed the outermost of the known planets, 4,528,000,000km from the Sun. The relative planetary positions are depicted.*

Should Pioneer 10 or 11 be recovered by an alien race, its point of origin will be revealed by this plaque.

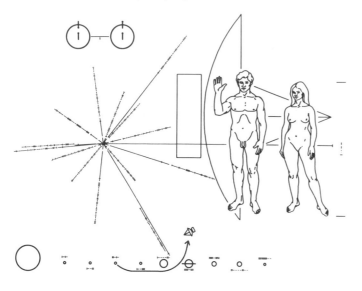

2.4 AU/year. Some time over the next few million years the inactive shell could be retrieved by intelligent beings out in the direction of Taurus. (It will have few stellar encounters, its closest approach being within 3.3 light years of the star Ross 248 in 33,000 years' time.) Should that happen, each Pioneer carries a 15 × 23cm gold-coated aluminium plaque etched with diagrams showing a man and woman, the solar system and its position relative to 14 pulsars. Distances are given in terms of the hyperfine transition line wavelength of hydrogen (21.11cm), a universal constant.

See Pioneer 11.

Venera 8

Launched: 0415 GMT March 27, 1972
Vehicle: A-2-e
Site: Tyuratam
Spacecraft mass: 1,184kg at launch (495kg capsule)
Destination: Venus
Mission: Lander capsule
Arrival: July 22, 1972 (landed 10°S/335° longitude at 0929 GMT)
Payload: Bus:
    Solar wind detector
    Cosmic ray detector
    Ultra-violet spectrometer
    + others?
  Capsule:
    Temperature and pressure sensors
    Anemometer
    Photometer (0.52–0.72μm)
    Gamma-ray spectrometer (0.3–3MeV)
    Gas analyser
End of mission: July 22, 1972
Notes: Fourth successful Venus capsule, second on surface, first day-side landing

Venera 8 is prepared for launch.

The successful landing of Venera 7 showed that the planet's surface pressure was "only" 90 Earth atmospheres, half of what the capsule had been built to withstand. Its successor was thus proofed to 105 atm and the weight thus saved devoted to extra thermal protection (this was the main limiting factor on the capsule's lifetime), additional instruments and a stronger parachute. A new top-cover release mechanism was incorporated and the parachutes were tested in wind tunnels with 500°C carbon dioxide to simulate Venusian conditions.

Venera 8 was the final flight by a first-generation probe – its twin, Kosmos 482, failed in Earth orbit four days later – before the much larger, more sophisticated second-generation models were introduced on the more powerful Proton booster. At 0542 GMT at 243sec injection burn from the 191 × 222km, 51.78° parking orbit sent Venera on its 482 million km Type I journey. A single course correction on April 6 ensured the first descent onto Venus' illuminated hemisphere. The target area was only 500km wide, dictated by the need for the Earth to be visible for data return and for a sufficiently steep entry angle to prevent the craft from skipping off the atmosphere and back into space.

The capsule's batteries were charged by the bus a few days before arrival and the refrigeration system took the internal temperature down to −15°C to prolong the surface life. The retaining straps were released at 0740 on July 22 (seemingly at an earlier stage than on Venera 7) and at 0837 the 495kg craft plunged into the atmosphere at 77° to the horizontal, travelling at 11.6km/sec. The fierce deceleration, at its maximum 67km high, slowed the descent to 250m/sec within 18sec. The main 'chute was deployed reefed initially and then fully opened at 30km.

The radar returns from the two antennae protruding over the side were used this time to estimate the density of the very top surface layer (the landing radars of US and USSR craft were used similarly on the Moon), yielding a figure of $0.8kg/m^3$. This indicated quite loose material, important to the design of new

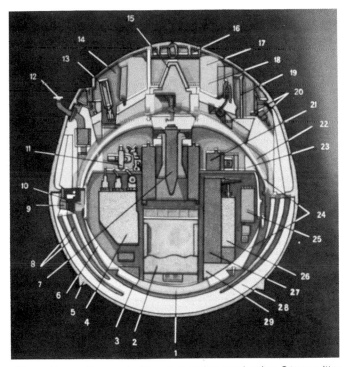

Above: Venera 8 capsule. **Key: 1** damping mechanism **2** transmitter **3** casing **4** commutator unit **5** aerodynamic damper **6** instrument frame **7** thermal system ventilator **8** air ducts **9** electrical connector **10** instrument frame **11** antenna system **12** bus cooling pipe **13** surface-deployed antenna **14** parachute section **15** antenna **16** cap **17** drogue chute **18** main chute **19** radio altimeter **20** cap ejection device **21** telemetry unit **22** heat-exchanger **23** heat accumulator **24** thermal protection **25** generator **26** commutator unit **27** timer and command unit **28** thermal protection **29** heat accumulator.

landers. Together with measurements of natural gamma radiation on the surface that indicated 4% potassium, $2 \times 10^{-4}\%$ uranium and $6.5 \times 10^{-4}\%$ thorium, this suggested a material similar to terrestrial granite. A similar spectrometer design was carried aboard the new-generation Venera 9/10 landers. The capsule came down in an upland plain region that is representative of 60–70% of the planet and is probably part of the ancient crust. More detailed analysis would be performed by X-ray fluorescence techniques on the new landers.

The atmosphere was again found to be 97% carbon dioxide, less than 2% nitrogen and less than 0.1% oxygen. But this time an additional instrument used a photosensor to observe the colour change in a chemical sensitive to ammonia: 0.1 and 0.01% were found at 46 and 33km high, respectively. Horizontal windspeeds were recorded by an anemometer: 100m/sec above 48km, 40–70m/sec between 42 and 48km and only 1m/sec below 10km, the general movement being in the direction of the planet's rotation. The high upper winds were confirmed by Mariner 10 in February 1974, when it discovered a four-day global circulation.

The new landers would be equipped to image the surface, so Venera 8 carried a simple cadmium sulphide photoresistor to measure the light flux. It found a sharp change at 30–35km – whereas most later probes indicated that the cloud deck ended at 48–49km – and extrapolation suggested that only 1.5% of the Sun's illumination reached the surface, resulting in a Stygian gloom similar to a terrestrial night just before dawn.

Veneras 9/10 accordingly carried floodlights, but the designers had been misled. Venera 8 had landed with the Sun only 5° above the horizon; noon-time was actually much brighter, equivalent to a dull, cloudy day on Earth.

The capsule touched down at 0929 and continued transmitting for another 50min. A second, omni, antenna had been released over the side to avoid a repetition of the Venera 7 communications problems. But although it was used for 20min, the main aerial on top proved adequate. The instruments now pinpointed ground-level conditions accurately: 470 ± 8°C and 90 ± 1.5 Earth atmospheres.

Though the first generation of Venus landers produced only basic data, they were essential and very successful precursors to the more adventurous new Veneras and the US atmospheric probes of December 1978.

Kosmos 482 Mass 1,184kg, launched March 31, 1972, by A-2-e from Tyuratam into 196 × 215km, 51.78° Earth parking orbit. The injection burn lasted for only 125sec and the final first-generation Venus probe was stranded in a 205 × 9,805km, 52.22° path. It re-entered during May 1981.

Apollo 16 Particles and Fields Subsatellite

Launched: 1754 GMT April 16, 1972 (ejected from Apollo 2207 GMT April 24)
Vehicle: Saturn V (No 511)/Apollo CSM-113
Site: ETR 39A
Spacecraft mass: 39kg
Destination: Moon
Mission: Orbiter
Arrival: April 19, 1972
Payload: Magnetometer
Particle telescopes (×2)
S-band transponder
End of mission: May 29, 1972 (lunar impact)

The Apollo 16 subsatellite, ejected from the Service Module experiment bay, was almost identical to the Apollo 15 model, the main difference being a magnetometer gain increased by a factor of two. Problems with the Apollo main engine forced the crew to release the subsatellite in a low, 100km 10° orbit. Planned mission lifetime was a year but the subsatellite impacted the surface after only 34 days (there was no propulsion system to raise perigee), although this generated unexpected low-altitude data. Tracking on the 416th revolution showed a perigee of 5km on the far side, and the subsatellite did not reappear to begin revolution 426, probably crashing around 10.2°N/112°E longitude at 2100 on May 29, 1972.

Experiment aims and results were similar to those for the Apollo 15 vehicle (qv).

Luna 21

Launched: 0655 GMT January 8, 1973
Vehicle: Proton (D 1-e)
Site: Tyuratam
Spacecraft mass: 5,800kg? at launch (840kg Lunokhod 2)
Destination: Moon

Lunokhod 2 mock-up at 1973 Paris Salon. (Graham Turnill)

Mission: Surface rover
Arrival: January 12, 1973 (landed 2235 GMT January 15 at 25.85°N/30.45°E)
Payload: Lunokhod 2:
  As Lunokhod 1 + visible/UV photometer, magnetometer, Rubin 1 photodetector, extra TV camera, modified Rifma X-ray spectrometer
End of mission: May ? 1973
Notes: Sixth Soviet lunar lander; second surface rover

Lunokhod 2 was an improved version of its predecessor, with more powerful electric motors permitting double the speed and an extra vidicon TV camera mounted high at the front giving the five-man steering crew working through the Yevpatoria station in the Crimea a better view as it returned a fresh image every 3sec. This time, the Lunokhod lid was opened during the journey to the Moon to charge the batteries but closed for major manoeuvres, including landing.

A 183 × 236km, 51.6° Earth parking orbit was followed by injection by the Proton final stage and a course correction on January 9. The main engine of the standard heavy Luna descent/insertion stage fired again on January 12 to establish a 90 × 100, 60°, 118min path around the Moon. Firings over the next two days produced a perigee of 16km, at which point the main engine reignited 255km from the target to brake the craft out of orbit after 40 revolutions.

The radar altimeter signalled another firing at a height of 750m until control was passed to the two smaller engines from 22m. These cut out 1½m up and the craft dropped on to the surface at 7km/hr at 2235 on January 15. It came to rest inside the rim of the 55km-diameter Lemonnier crater (a subdued flooded ring), just 180km north of the Apollo 17 site on the eastern edge of the Sea of Serenity. Apollo 17 had concluded the first phase of manned exploration only a month earlier.

The cameras inspected their surroundings and at 0114 the next day Lunokhod 2 rolled down the ramp and parked some 30m away for two days to charge its batteries. This rover was not as long-lived as its predecessor, but in five lunar days it

traversed 37km (peaking at 16½km on March 11–23), returned 80,000 TV pictures and 86 panoramas, carried out at least 740 soil mechanical tests, and was struck by laser beams from the Crimean and French Pic du Midi observatories more than 4,000 times. The Rubin 1 device acknowledged each contact with a radio signal and the Earth–Moon distance was determined to within 20–30cm.

The controllers were more adventurous this time as they piloted the rover among the Taurus Mountain foothills of the region. On one occasion they steered Lunokhod 2 up an 18° slope and 100m into a crater, only to find the wheels sinking in up to the axles. They backed it up and departed by a safer route. A 300m-wide, 16km-long fissure was discovered and thoroughly investigated during the fourth lunar day (April 8–23), metre-sized boulders close to the edge making navigation difficult. The new magnetometer on the end of a 2.5m boom at the front showed a variation in the magnetic field near the rim, and an important conclusion from the mission was that the Moon has a weak permanent magnetic field (in agreement with Apollo's findings). A thermometer with the magnetometer recorded a temperature of −183°C during the first lunar night.

The astrophotometer unexpectedly showed the sky to be 13–15 times brighter than the Earth night sky, throwing doubt on the operation of future lunar observatories in visible and UV light. The improved Rifma-M X-ray fluorescence spectrometer discovered that the elemental iron content of the soil fell as Lunokhod 2 ventured into the hills. Close to the lander it found 24 ± 4% silicon, 8% calcium, 6% iron and 9% aluminium, whereas Lunokhod 1 had registered 10 20% iron.

It is not known when activities ended, but it seems likely that a breakdown in mid-May prevented a final parking manoeuvre to position the laser reflectors for continued use. No further rovers have been flown.

See Luna 17/Lunokhod 1.

Pioneer 11

Launched: 0211 GMT April 6, 1973
Vehicle: Atlas-Centaur 30 + TE-M-364-4 (Atlas No 5011D)
Site: ETR 36B
Spacecraft mass: 258.5kg at launch
Destination: Jupiter, Saturn, interstellar space
Mission: Flyby
Arrival at target: Jupiter: December 3, 1974, closest approach 42,960km at 0522 GMT
  Saturn: September 1, 1979, closest approach 20,900km at 1631 GMT
Payload: As Pioneer 10 + fluxgate magnetometer
End of mission: 1994?
Notes: First Saturn flyby, discovery of F-ring, second jovian flyby

Pioneer 11, almost identical to its predecessor, was allowed to venture three times closer to Jupiter but, because it was travelling much faster (172,000km/hr), it suffered less radiation damage. Following the direct-ascent trajectory launch, a burn with Pioneer's own hydrazine thrusters on April 11, 1973, altered the Type I path to permit the Saturn flyby option. Pioneer 11 would approach Jupiter from the south, whip northwards and out 1 AU above the ecliptic by May 1976 before descending to meet Saturn.

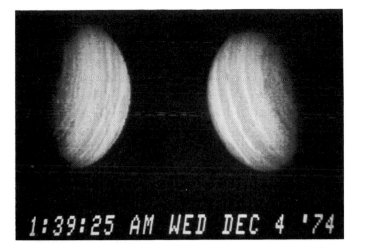

Left: Jupiter seen from 2.1 million km above the north pole. The Great Red Spot is visible near the equator.

Below left: *One of the best overall views of Saturn by Pioneer 11, taken from 2½ million km on August 29, 1979. Satellite Rhea is included in the frame.*

atmosphere above them. They were also several degrees colder. Many small-scale convective cells were spotted and in the Great Red Spot itself features suggesting convection and circulation were evident.

Pioneer survived the Jovian encounter in general good health, but the meteoroid detector began to generate spurious signals interfering with other instruments and had to be turned off on April 16, 1975. The plasma detector refused to revive after a dormant period but it was eventually brought back on line December 3, 1977. The probe had travelled a total of 3,200 million km as it approached Saturn at an angle of 6.55° to the equator on September 1, 1979. The scientists had decided in May 1978 to guide it past outside the rings – as Voyager would be – instead of taking the exciting but risky route through the gap between the inner rings and the planet.

The instruments detected the bow shock on August 31, 1979, some 1,445,750km out, providing the first conclusive evidence of a magnetic field. Pioneer 11 crossed the ring plane for the first time – the feared battering from dust did not materialise – at 1436 on September 1 and headed for a closest approach of 20,900km, its speed rising to 114,080km/hr. Two hours later it crossed the rings in the opposite direction on the way out.

Perhaps the most exciting discovery was a narrow ring outside the A ring, dubbed "F". Scientists also spotted a new satellite 200km in diameter and were horrified to find later that Pioneer 11 had missed it by only a few thousand kilometres (it later emerged that this was one of two co-orbiting moons seen uncertainly in 1966). Saturn's overall temperature was measured at −180°C, indicating that it is emitting 2½ times as much heat as it receives from the Sun.

The atmosphere was more featureless than Jupiter's and presented a somewhat bland face to the unsophisticated imaging system. The magnetic field had a dipole moment 540 times greater than Earth's and, surprisingly, it coincided with the planet's rotational axis. Energetic charged particles were found to be absorbed by the inner satellites and rings, to the extent that intensities were several hundred times less than at Jupiter. Calculations based on Pioneer 11 results suggested that Saturn is primarily liquid hydrogen with a core of about 10 Earth masses.

Pioneer 11 is now leaving the solar system and heading towards the centre of the Galaxy in the general direction of Sagittarius. In 1990 it will cross Neptune's orbit, followed by Pluto's during the next year, and continue starwards at 2.2 AU/year.

See Pioneer 10 for spacecraft description and interstellar mission details.

The probe entered the Asteroid Belt on March 20, 1974, but registered only seven micrometeoroid strikes during its seven-month traverse, confirming Pioneer 10's finding that the dust hazard had been overestimated. A 64.07m/sec burn on April 19 lasting for 42min 36sec and consuming 7.7kg of hydrazine shifted the planned closest approach to within 42,960km of Jupiter's cloud tops.

The far-encounter phase began on November 7, 1974, as Pioneer crossed the orbit of outermost known moon Sinope, and on November 25 the planet's bow shock was penetrated. Scientists were surprised to see the shock slip past again two days later in the other direction as the Sun became more active, allowing Pioneer to spend a further 5½hr outside the magnetosphere. The huge, balloon-like magnetic field was being pushed around by the fluctuating solar wind and the Sun's own magnetic field. Pioneer 11 found Jupiter's field to be somewhat more complex than its predecessor had indicated. The latest probe carried a fluxgate magnetometer mounted on its main body to allow more precise measurements at higher levels during closest approach.

The craft slid into occultation at 0502 GMT on December 3, followed by the closest approach 20min later. It emerged at 0544, the signals taking 40min to reach Earth. These first views of the north polar regions showed the cloud tops to be substantially lower than those at the equator, with a thicker layer of

Explorer 49 (Radio Astronomy Explorer 2) Mass 330kg, launched on direct ascent at 1414 GMT June 10, 1973, by Delta No 95 from ETR 17B. Remains last US lunar mission, although it used the Moon only to block terrestrial radio interference and permit sensitive radio astronomical measurements of the Sun, Jupiter and deep space. As such, it is not described in detail here. The 92cm-diameter, 79cm-high

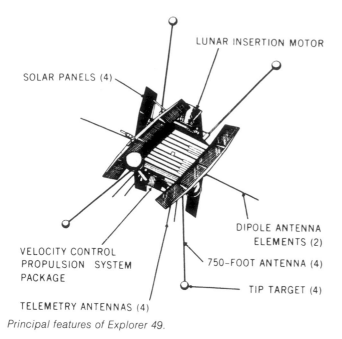

Principal features of Explorer 49.

vided the best window for 15 years, but two years later not even the powerful Proton booster could inject a combined orbiter/lander towards the planet. The vehicle had to be divided into less massive flyby/landers (Mars 4 and 5) and orbiters (Mars 6 and 7) to achieve the same results. Mars 6 and 7 were virtually identical to the Mars 2/3 vehicles except that the propellant load of almost two tonnes had to be reduced to about 300kg (estimates by Clark). Mars 4 and 5, while carrying the full propellant load to brake into Mars orbit, had to do without the landing capsules and carry experiment platforms in their place.

Mars 4 was the first away, completing an orbit correction on July 30. But as it approached the planet the main engine failed to fire and it was left to sail past 2,200km above the surface on February 10, 1974. Nevertheless, the imaging system did return some pictures (see Mars 5). It is not clear if the Mars 2/3 method was used, scanning exposed photographs to produce high-quality results, or if a TV-vidicon was carried. The orbiter should have acted as a relay for the Mars 6 lander, and the failure meant that Mars 5 had to be used for this purpose instead.

aluminium truncated cylindrical main body carried four fixed paddles, each with 3,520 solar cells to supply a total of 38.5W to six nickel-cadmium batteries. Once in lunar orbit, the solid braking motor was ejected (reducing satellite mass to 200kg) and attitude-control and orbit adjustments were performed by hydrazine thrusters. Four 5cm-wide, 229m-long beryllium/copper radio antennae unwound to create a sensitive receiver measuring 457m tip-to-tip. Two slow-scan TV cameras (13sec/frame) with 1.27cm vidicons were carried on the solar panels to correlate radio signals with optical sources. Two tape recorders collected data over the lunar far side for subsequent playback to Earth.

A course correction en route to the Moon was made at 1528 on June 11 (a planned second burn was not required) and the insertion burn was carried out at 0721 on June 15 to create a 1,123 × 1,334km, 61.3° lunar orbit. The braking motor was jettisoned and the hydrazine system established a 1,053 × 1,063km, 38.7°, 221min path. The two antennae extended to only 183m but the mission was not affected. NASA judged the mission to have been successfully completed in June 1975. Transmissions were terminated in August 1977.

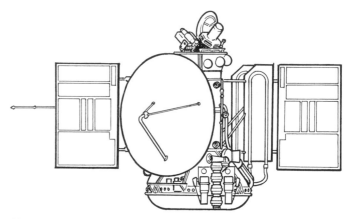

Mars 4/5 orbiter, with the main scientific payload at top. (David Woods)

Mars 5

Launched: 1856 GMT July 25, 1973
Vehicle: Proton (D-1-e)
Site: Tyuratam
Spacecraft mass: 4,385kg? at launch
Destination: Mars
Mission: Orbiter
Arrival: February 12, 1974
Payload: TV system (52mm and 350mm focal length)
Infra-red radiometer (8–20μm)
Ultra-violet photometer (2,600Å, for ozone measurements)
5 photometers (2μm CO_2 band, 1.38μm H_2O band, 0.3–0.8μm, 2–5μm, 1,215Å)
3.5cm radio receiver (temperature and surface density)
2 polarimeters (0.35–0.8μm in 9 bands)
Gamma-ray spectrometer
Magnetometer
Charged-particle detectors
End of mission: ?
Notes: Fourth Mars orbiter (third Soviet)

Mars 4

Launched: 1931 GMT July 21, 1973
Vehicle: Proton (D-1-e)
Site: Tyuratam
Spacecraft mass: 4,385kg? at launch
Destination: Mars
Mission: Orbiter
Arrival: February 10, 1974
Payload: Similar to Mars 5
End of mission: Failure of main mission February 10, 1974; date of last contact unknown

While the US delayed its Viking orbiters/landers to 1975 for financial reasons, the Soviets chose to launch no fewer than four Mars craft during the 1973 opportunity. 1971 had pro-

Mars 5 was the only fully successful member of the four-vehicle fleet dispatched to the planet in July/August 1973. It was injected into a Type I transfer path at 2015 on July 25 and performed a trajectory correction with its main engine on August 3. The engine flared into life again at 1545 on February 12, 1974, to brake Mars 5 into a 1,760 × 32,560km, 35.33°, 24hr 53min orbit.

The spacecraft acted briefly as a communications relay for the Mars 6 lander, and together with Mars 4 returned 60 images of the surface. Their quality appears comparable with those of Mariner 9 in 1972–2, exhibiting features that suggest erosion caused by free-flowing water. Other instruments located an ozone layer 30km above the surface, constituting $10^{-5}\%$ of the atmosphere by volume, and revealed that the outermost region of the atmosphere is atomic hydrogen extending 20,000km above the surface.

See Mars 4 for spacecraft details.

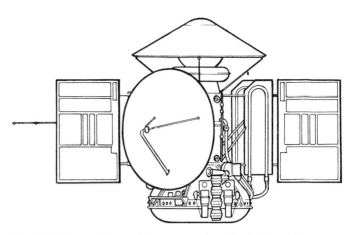

Mars 6/7 flyby module and descent section. (David Woods)

Mars 6

Launched: 1746 GMT August 5, 1973
Vehicle: Proton (D-1-e)
Site: Tyuratam
Spacecraft mass: 3,495kg? at launch
Destination: Mars
Mission: Flyby/lander
Arrival: March 12, 1974 (landed at 24°S/25°W)
Payload: Flyby module:
    Magnetometer
    Solar wind detector
    Micrometeorite detector
    Cosmic ray detector
    French Stereo experiment (see Mars 3)

Mars 5 image of a 25km-diameter Martian crater.

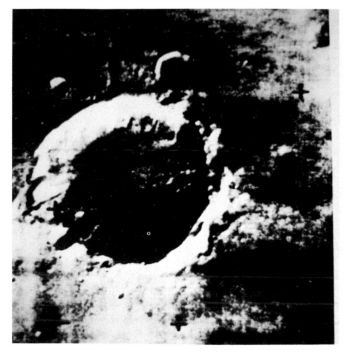

Lander:
    TV system
    Mass spectrometer
    Temperature and pressure sensors
    Soil chemical/mechanical analyser
End of mission: ?
Notes: Third object on Mars

Mars 6 was the last of the four Soviet probes to approach the Red Planet in 1974, having followed a slightly slower path than its Mars 7 twin. A trajectory correction had been executed by the main engine on August 13, 1973.

At 0502 on March 12, 1974, still 55,000km out, the flyby bus ejected the lander section before flying past the planet at 16,000km. At 0906 the capsule (see Mars 2/3) hit the atmosphere at a speed of 5.6km/sec, its heatshield reaching a temperature of 1,000°C, and began its automatic sequence of parachute deployment. The speed had been reduced to 600m/sec by 090832. Contact was maintained for 148sec following main 'chute opening, but just a few seconds before the anticipated landing the signal was lost. However, data relayed back during the descent indicated surface pressure and temperature of ~6mb and −43°C respectively. The most intriguing information came indirectly from the mass spectrometer pump sampling the atmosphere: it indicated that 20–30% was an inert gas, probably argon, a result viewed with suspicion by US scientists and subsequently disproved by Viking – though not before the US landers' analyser designs were modified.

Mars 7

Launched: 1700 GMT August 9, 1973
Vehicle: Proton (D-1-e)
Site: Tyuratam
Spacecraft mass: 3,495kg? at launch
Destination: Mars
Mission: Flyby/lander
Arrival: March 9, 1974 (lander missed)
Payload: As Mars 6
End of mission: ?

Mars 7 was the third failure in the four Mars attempts of 1973–4. A trajectory correction was completed successfully on August 16, 1973, and the lander section was released as planned on March 9, 1974. Unfortunately, it missed the planet by 1,300km, indicating that its solid-propellant motor or attitude control system (see Mars 2/3) had failed in some manner.

This final fleet brought Soviet attempts on Mars to a close for 15 years. Possibly other missions were planned but the outstanding achievements of the US Vikings and their own dismal record persuaded the Soviets to concentrate on Venus. The last of this generation of Mars/Venus probes flew as the Vega Halley's Comet/Venus craft of 1984–6, the announcement in 1985 of the Mars-Phobos orbiter/landers for launch in 1988 marking the introduction of a new type.

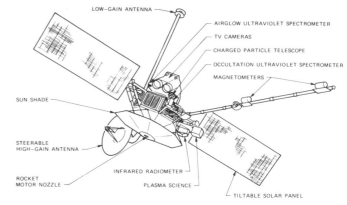

Principal features of Mariner 10.

Mariner 10

Launched: 0545 GMT November 3, 1973
Vehicle: Atlas-Centaur 34 (Atlas No 5014D) + TE-M-364-4
Site: ETR 36B
Spacecraft mass: 502.9kg
Destination: Venus, Mercury
Mission: Flyby
Arrival at target:
   Venus: February 5, 1974, closest 5,768km at 1701 GMT
   Mercury 1: March 29, 1974, closest 703km at 2047 GMT
          2: September 21, 1974, closest 48,069km
             at 2059 GMT
          3: March 16, 1975, closest 327km at 2239 GMT
Payload: TV system
Infra-red radiometer
Ultra-violet airglow spectrometer
Ultra-violet occultation spectrometer
Magnetometers (2)
Charged-particle telescope
Plasma analyser
End of mission: March 24, 1975
Notes: First multiple planetary swingbys, first Mercury encounter

Very little was known of the planet Mercury until the US Mariner 10, approved by NASA in 1969, completed three flybys during a one-year period in 1974–5. In order to hold down costs, the US agency employed a single spacecraft (most planetary missions were conducted in pairs) launched by an Atlas booster instead of the more expensive Titan III. The latter could have flung a payload onto a direct Mercurian trajectory, but the smaller, cheaper vehicle was usuable when windows to Mercury via a Venus swingby opened in 1970 and 1973. Analysis showed that the second opportunity would permit Mercury re-encounters at six-monthly intervals, and so the window of October 16–November 21, 1973, was adopted. But this was no easy option: if Mariner 10 was to reach Mercury, the gravitational assist at Venus would have to be controlled to within an unprecedented 400km by several accurate mid-course corrections.

Mariner 10 was a typical three-axis-controlled spacecraft of the type that had seen extensive service on Mars missions. The 18kg magnesium octagonal main body was 46cm high

and 138cm in diameter, centred on a tank holding 29kg of hydrazine for the 220N restartable engine that protruded through the face of the deployable sunshield. The enlarged tank capacity compared with previous Mariners would permit a total ΔV of 122m/sec. Since Mariner would be venturing so close to the Sun, thermal control was vital; the sunshield, plus louvres and insulating blankets, would maintain the electronics within their operating temperature ranges. Likewise, the two 269 × 97.5cm solar panels with 19,800 cells to produce 820W at encounter were rotatable about their long axes to maintain their temperature at 115°C.

The roll and yaw jets of the attitude-control system (with 3.6kg of nitrogen supply) were mounted at the solar-panel tips; the pitch thrusters were housed on the main body. The two 1,500mm and 62mm f/8.5 TV cameras, each with an 8-position filter wheel and a 9.8 × 12.3mm vidicon surface to generate 700-line pictures, and the UV airglow spectrometer were mounted on a scan platform on one face of the main body. The two magnetometers had their own 6m boom, while the rest of the total of 78kg of instruments used the main structure. The 119cm-diameter aluminium honeycomb articulated high-gain antenna would transmit the 117,600 bits/sec generated by the cameras at encounter.

Mariner 10 was accelerated to a speed of 40,962km/hr following a dual Centaur burn that included a 188km parking orbit. Some 16¼hr after launch the first TV pictures, of Earth and then the Moon, were returned to check the cameras. In fact Mariner 10 provided the very first clear views of some northern lunar areas. But after this promising start ground engineers had to nurse the spacecraft throughout the mission as a series of problems arose. The plasma analyser door jammed partially open and the TV heaters failed to switch on, requiring the cameras to be kept on to maintain a working temperature. Problems with the high-gain antenna feed emerged during Christmas 1973, and on January 28, 1974, the attitude-control system wasted 16% of the nitrogen supply following a gyro malfunction. A switch was made from inertial to Sun and Canopus control for the imminent Venus encounter to avoid possible difficulties. On January 8 spacecraft power had been mysteriously switched irreversibly to the back-up system, leaving no margin for future error.

Launch had been accurate but a 19.9sec, 7.77m/sec burn using 1.8kg of hydrazine on November 13, 1973 was necessary to adjust the Venus arrival point to avoid missing Mercury completely. As expected, a second firing of the 220N engine was needed because the first had left Mariner 10 still 1,384km too far out at Venus and arriving 2min too early. A 3.8sec, 1.2m/sec burn on January 21, 1974, took the trajectory to within 27km of the perfect aim point.

Mariner 10 approached Venus from the night side and returned the first pictures – showing the lighted cusp at the north pole – at 1650 GMT on February 5. Twelve minutes later the spacecraft dipped to its nearest point above the dense clouds, and six minutes after that dropped into occultation behind the limb. The radio signals continued for several minutes as they traversed the atmosphere, probing the various cloud layers.

The most interesting pictures came in as the craft left Venus behind. The earlier sequences showed only bland clouds but the cameras now used their UV filters and revealed a wealth of detail: great bands encircled the globe, providing scientists with data on atmospheric circulation. The upper atmosphere took just four days to complete a full rotation while the planet itself required 243 Earth days. Subsequent US and Soviet atmospheric probes confirmed that the 300km/hr windspeeds implied by Mariner were real. The infra-red radiometer showed the cloud tops to be at −23°C on both the day and night sides, while the fields and particles instruments returned information on the solar wind interactions with the ionosphere (Venus has almost no magnetic field). The final TV image, on February 13, brought the total to 4,165.

Below: *Mariner 10's flyby of Venus produced this ultra-violet image showing upper-atmosphere circulation. It was taken at a distance of 760,000km the day after closest approach.*

Right: *Mosaic of 18 images taken about six hours after closest approach during the first Mercurian encounter. At centre left is part of the Caloris Basin. North is at top.*

The faulty gyros ruled out one of two course corrections en route to Mercury, but a 51sec, 18m/sec burn on March 16 provided a flyby distance that was acceptable though 200km closer than originally planned. The first images of Mercury, taken on March 23 some 5.3 million km out, were no better than Earth-based views. But features began to appear as the distance reduced, with vague white spots gradually resolving into very light craters. One – named after prominent planetary astronomer Gerald Kuiper, who died during the mission – was especially notable. It became clear that Mercury was astonishingly Moon-like. Craters, ridges, lava-flooded areas, and chaotic and jumbled terrain all appeared.

A 1,300km-diameter mountain ring outlined a huge meteorite impact basin, since named the Caloris Basin, looking rather like an eye on Mercury's side. It was difficult to distinguish between these new pictures and the Ranger or Lunar Orbiter photographs of the Moon.

In a complete surprise the magnetometers revealed a magnetic field 1/60th as strong as Earth's; its origin is still a

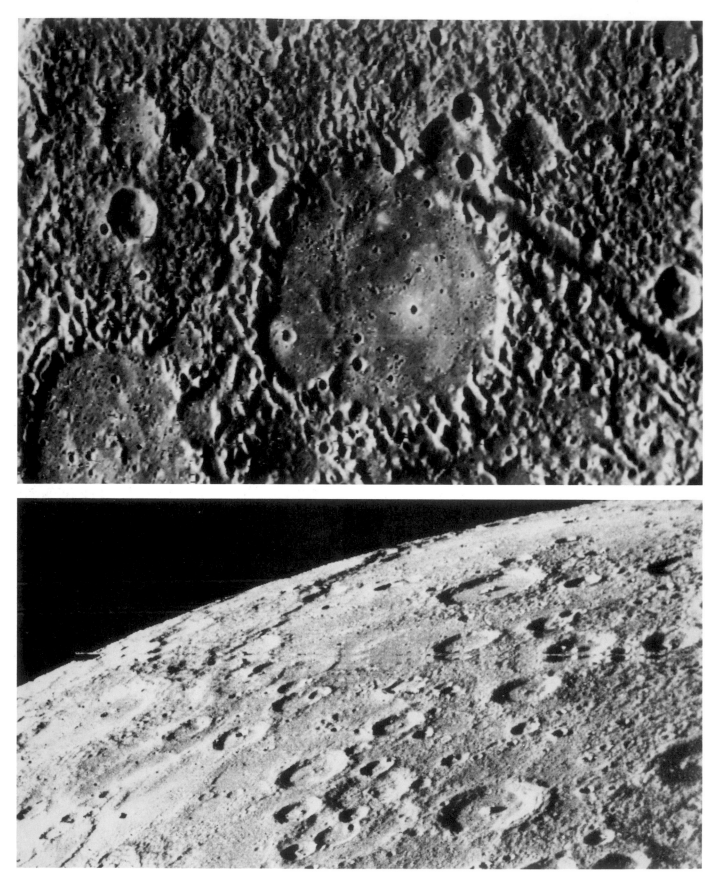

Left: The jumbled terrain antipodal to the Caloris Basin may have been created by the focusing of seismic waves from that large impact structure's birth. This image was produced during the first encounter, in March 1974.

Below left: The scarp near the top of this March 1974 encounter image is about 3km high and runs for hundreds of kilometres. Such features, not found on the Moon, appear to have been caused by the planet's crust shrinking around its large metallic core.

mystery. Mercury's density of 5.5g/cc, derived from tracking Mariner 10, is almost the same as Earth's, so there must be a metallic core containing about 80% of the total mass. The radiometer put the night-time temperature of the pre-dawn surface at −183°C, the maximum daytime temperature at 187°C. The TV also searched for Mercurian moons but, though there was a false alarm caused by a star, none were revealed.

As Mariner 10 looped around the Sun, three further course corrections – on May 9 (50m/sec), May 10 (27.7m/sec) and July 2 (3.35m/sec) – produced a second but more distant flyby on September 21. Solar pressure on the panels and high-gain dish was used for attitude control in order to conserve the diminishing nitrogen supply for a possible third encounter. Mariner 10 passed over the same hemisphere again, its more southerly route this time exposing new regions. The flyby was too remote for most of the instruments but geologists benefited from fresh perspectives with the TV system. The UV spectrometer set the upper limit on the planet's helium atmosphere – acquired from the solar wind – at 10^{-15} Earth density.

Mariner 10 returned to Mercury on March 16, 1975, following course corrections on October 30, 1974, February 13, 1975 and March 7, 1975. This, the closest encounter of all, gave the magnetometers an opportunity to delve into Mercury's unexpected field. The 349 images produced on this pass displayed resolutions down to 140m.

The nitrogen supply finally ran out on March 24 and the transmitters were commanded to shut down. Despite its almost continuous problems, Mariner 10 completed one of the most successful planetary explorations of all: four encounters in 13 months. Its cameras imaged only about half of the Mercurian surface so, bearing in mind the surprises produced by Mariner 9 orbiting Mars, there is plenty of scope for new discoveries. There are however no plans to return to the planet in the foreseeable future.

Luna 22

Launched: 0857 GMT May 29, 1974
Vehicle: Proton (D-1-e)
Site: Tyuratam
Spacecraft mass: 5,700kg at launch (~1,900kg in lunar orbit)
Destination: Moon
Mission: Orbiter
Arrival: June 2, 1974
Payload: As Luna 19
End of mission: September 2, 1975
Notes: Sixth Soviet lunar orbital survey (second by third-generation craft)

The progression of Luna missions at roughly yearly intervals continued when Luna 22 was fired into a 178 × 227km, 51.6°

Earth parking orbit and then dispatched Moonwards using the Proton's escape stage. The single course correction by the main engine was executed the following day, and astronomers at the Georgia Astrophysical Observatory acquired the vehicle some 250,000km out from Earth. The new Lunas usually made two trajectory corrections, but perhaps on this occasion the launch had been particularly accurate.

There had been a profusion of samplers and rovers but only one third-generation orbiter before the new vehicle braked into a circular 220km, 19.6°, 130min orbit on June 2. Apparently identical to Luna 19, the new orbiter adjusted its path on June 9 to 25 × 244km, 120min to swoop in close to the surface for the photographic phase of the mission, possibly returning high-resolution images of candidate landing sites for Luna 23. Pictures from greater altitude were also acquired to cover larger areas at reduced resolution. No photographs from the mission have been released. The altimeter and gamma-ray spectrometer also benefited from the low orbit, producing detailed information on surface topography and elemental composition respectively.

The photographic phase over, the orbit was boosted to 181 × 299km on June 13. The three-axis control may also have been switched to spin stabilisation to conserve propellant for a lengthy fields and particles survey. A primary mission objective was the mapping of variations in the gravitational field, and no major manoeuvres were executed for almost five months in order to reveal orbital changes. A total of 23 meteoroid impacts of greater than 10^{-11}g were detected during June–August 1974, a rate three times lower than was found in the dipping imaging phase. On November 11, 1974, shortly after Luna 23 had entered orbit and then descended to the surface, the orbit was made more eccentric: 171 × 1,437km, 19.55°, 192min. This new path sliced through a larger region for more comprehensive measurements around the Moon.

On April 2, 1975, having completed 2,842 revolutions, Luna 22 moved to a 200 × 1,409km, 21.0°, 192min orbit, where it remained for the first anniversary of orbital insertion. The Soviets announced that Luna 22 was still operating after 3,296 revolutions and 2,175 communications sessions but that its main programme had ended. This was underlined on August 24, when a burn took the craft to within 30km of the surface (no apogee was given) and the camera returned a picture to demonstrate the equipment's reliability after a year's inactivity.

A few days later the orbit was raised to 100 × 1,286km, 21.0°, possibly to extend Luna 22's lifetime as the propellant tanks began to run dry. Manoeuvres since attaining orbit had amounted to a ΔV of about 350m/sec. The "manoeuvring propellant" – presumably meaning the attitude-control gas essential to stabilisation – was reported to have been consumed by September 2, although there was a report the mission was not finally over until November. Some 30,000 commands had been received from Earth by the time of Luna 22's demise.

Luna 22 remains the last dedicated lunar orbiter from any nation. None of the Soviet craft strayed far from the equatorial regions, though there is a declared intention to fly a new-generation polar orbiter in the early 1990s for global geochemical mapping. Part of its task will be to search for water-ice pockets near the eternally shadowed poles, a vital step towards establishing a lunar base and a viable transport system using indigenous oxygen and hydrogen propellants.

See Luna 19.

Luna 23

Launched: 1430 GMT October 28, 1974
Vehicle: Proton (D-1-e)
Site: Tyuratam
Spacecraft mass: 5,700kg? at launch (1,880kg? on surface)
Destination: Moon
Mission: Sample return
Arrival: November 2, 1974
Payload: As Lunas 16/20, with improved sample drill
End of mission: November 9, 1974 (contact ended)
Notes: Fifth Soviet lunar lander

The exciting series of Luna missions, broken only by the Luna 18 crash, appeared to be progressing smoothly as Luna 23 headed for a soft landing at around 13°N/62°E in the SE region of Mare Crisium on November 6, 1974. Sample-return missions were limited to longitudes around 60°E in order to avoid horizontal velocity demands on the simple ascent vehicle. At the same time, this line of longitude fortuitously runs through the previously uninvestigated Mare Crisium. The successful sample-return flights of Lunas 16 and 20 had drilled down 30cm at most, but the new craft was equipped with an improved device to reach a depth of 2m (see Luna 24).

A 182 × 246km, 51.5° Earth parking orbit was followed by trans-lunar injection by the Proton's final stage and a course correction on October 31. At around 2150 on November 1 Luna 23's main engine fired to brake the craft into a 94 × 104km,

Artist's impression of a Luna 23/24 craft with the new type of drilling mechanism.

138°, 117min orbit. Adjustments on November 4 and 5 established a pre-descent path of 17 × 105km and, after completing about 50 revolutions since arriving, Luna 23 headed down to an 0537 landing.

On a normal mission the TV system would have looked for a suitable drilling position in preparation for sending a sample to Earth after about a day of surface activities. Arrival in the Soviet Union would have occurred on November 10 or 11. Unfortunately, a rough landing – the Soviets ascribed it to the nature of the terrain – damaged the drill and the main mission had to be abandoned. Possibly the surface had confused the radar altimeter and the main engine had started too late (600m on Luna 16) or had cut out too high. A similar problem had affected Surveyor 3, but in that case the descent engines had burned for too long and the craft made several hops following initial contact.

A damaging impact is indicated by the fact that the ascent stage and capsule were not launched simply for the experience. Whatever the reason, ground controllers could do nothing and contact was broken on November 9. The site was clearly considered to be important because a successful attempt was made by Luna 24 in August 1976.

See Lunas 16, 20 and 24.

Venera 9

Launched: 0237 GMT June 8, 1975
Vehicle: Proton (D-1-e)
Site: Tyuratam
Spacecraft mass: 4,936kg at launch (1,560kg descent section, including 660kg lander; orbiter 2,283kg plus 1,093kg propellant)
Destination: Venus
Mission: Orbiter/lander
Arrival: October 22, 1975 (landed 0513 at 32°N/291° longitude)
Payload: Bus:
        Imaging system
        UV imaging spectrometer (French-supplied)
        IR radiometer (8–30μm) and spectrometer (1.5–3μm)
        Photopolarimeter (4,000–7,000Å)
        Magnetometer
        Ion/electron detectors
        Optical spectrometer
    Lander:
        Panoramic imaging system
        Mass spectrometer
        Temperature and pressure sensors
        Anemometer
        Nephelometer
        Photometers
        Gamma-ray spectrometer
        Radiation densitometer
        Accelerometers
End of mission: October 22, 1975 (lander) and March 1976? (orbiter)
Notes: First Venus orbiter and surface picture

The advent of the second-generation Venera craft began a new era in Soviet exploration of the planet. Sophisticated landers and instrumented orbiters could now undertake

Above: Venera 9/10 craft, with lander at left (note floodlights for illuminating surface).

Below: Venera 9/10. **Key: 1** orbiter **2** descent sphere **3/8/14** scientific instruments **4** high-gain antenna **5** propellant tanks **6** radiator **7** Earth sensor **9** Canopus (star) sensor **10** Sun sensor **11** omni antenna **12** instrument bay **14** attitude-control gas tank **15** radiator **16** attitude-control thrusters **17** magnetometer **18** solar panels.

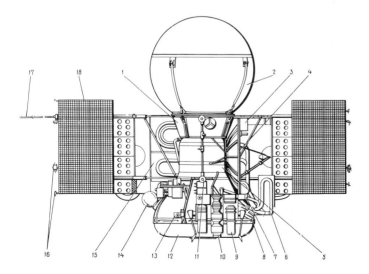

lengthy surveys to build a comprehensive picture of Venus and its environs. The A-2-e launcher was finally retired from planetary service and the more powerful Proton D-1-e, long since used for Luna (1969) and Mars (1971), was brought on line.

The orbiter/flyby bus was similar to that already in operation with the Mars programme, with a height of 2.8m and solar-panel span of 6.7m. The central 1.1m-diameter, 1m-high cylinder housed the propellant tanks for the KTDU-425A 9.86–18.89kN main engine, which could be ignited up to seven times. The nozzle fired through the centre of a 2.35m conical-toroidal pressurised instrument section at the bottom.

Viewing over the rim and along the same axis were Sun and star sensors to provide reference information for the three-axis attitude-control system. Gyros for inertial reference during manoeuvres were carried inside, where a digital computer made the new craft less reliant on direct Earth commands. Inboard of the 1.25 × 2.1m solar panels were radiators designed to dump excess heat from the equipment section via a cooling fluid that could also be diverted up to the descent module through external pipes.

Although the bus' array of photometers, spectrometers and radiometers would survey the upper atmosphere, and the magnetometer and radiation detectors would investigate near-Venus conditions, the craft's major role was to act as a relay for the lander's data. The orbit (or flyby on later missions) was arranged so that Venera 9 would be within sight of the surface capsule, allowing it to receive the capsule's data and

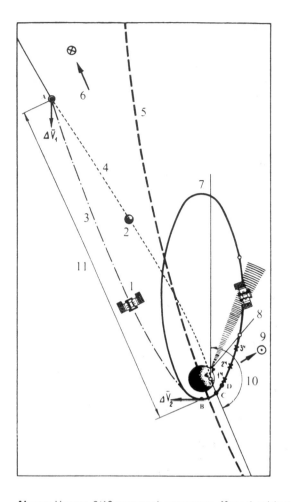

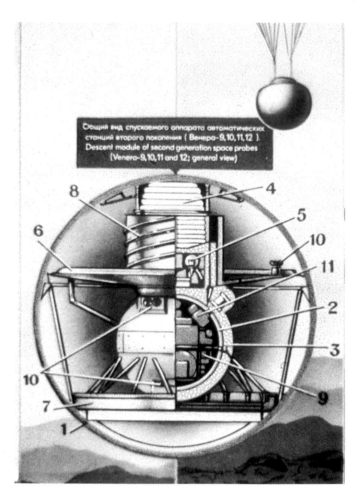

Above: *Venera 9/10 approach sequence.* **Key: 1** *orbiter* **2** *descent sphere* **3/4** *approach hyperbola* **5** *orbit of Venus* **6** *Earth direction* **7** *orbiter's path* **8** *lander's radio horizon (ie, Earth out of direct contact)* **9** *solar direction* **10** *lander-orbiter radio contact zone* **11** *automatic flight sequence (48hr):* A *sphere separation and manoeuvre by bus* B *bus braking burn* C *beginning of radio visibility* D *start of orbiter contact with lander.*

Above right: *Cutaway of the descent sphere.* **Key: 1** *descent sphere with ablative coating* **2** *pressurised container* **3** *thermal insulation* **4** *parachutes* **5** *scientific instruments for descent* **6** *aerodisc* **7** *lander ring* **8** *antenna* **9–11** *scientific instruments (11 is clearly the camera).*

store them on tape for subsequent transmission to Earth over its more powerful radio system. Earlier Veneras had descended over the night side – Venera 8 was only just into daylight – where they could transmit directly to Earth.

An Earth parking orbit of 171 × 196km, 51.54° was followed by injection into a Type I trajectory and two course corrections, on June 16 (12.0m/sec) and October 15 (13.5m/sec). The 5m-high craft charged the lander's batteries and cooled its interior to −10°C in preparation for release of the 2.4m sphere on October 20. The descent section headed for the daylight hemisphere while the bus corrected its path with a 247.3m/sec burn on October 20 to travel around Venus the other way. A 922.7m/sec braking burn created an orbit of 1,500 × 111,700km, later corrected to 1,300 × 112,200km,

48hr 18min and then 1,510 × 112,200km, 34.2°, 48hr 18min. The orbiter headed out towards its first apogee and recorded the lander's signals as it did so.

The descent sphere hit the atmosphere some 125km high at 0358 on October 22. It was travelling at 10.7km/sec and 20.5° to the horizontal, a shallower entry than those of previous Veneras. This produced gentler deceleration, and accelerometer telemetry was returned twice a second to characterise the atmosphere over 100–63km. Once the aerothermal sphere had done its job and slowed the craft to 250m/sec 64km above the surface, the upper portion was jettisoned and a 2.8 braking parachute and then three 4.3m main 'chutes were deployed. The lower portion was then released to leave the craft descending to the surface with instruments operating. It emerged from the cloud deck at a height of 50km and released the canopies to free-fall to the surface. The mass spectrometer was sampling the atmosphere and the nephelometer and photometer recorded the extent and particle sizes of the cloud strata.

The central 1m-diameter pressure vessel containing the electronics of the command, telecommunications and experiment systems was attached by shock-absorbers to a base "doughnut" that could partially collapse to absorb the 7–8m/sec impact. Atmospheric gas filled the ring through holes to enhance the cushioning effect. Above was a 2.1m-diameter aerobraking disc that also served as the antenna reflector; the earlier Veneras had demonstrated that parachutes were not necessary to reach the surface safely because the atmosphere is so dense. Above the disc a helical antenna was wound around the 80cm-diameter, 40cm-high cylinder that had housed the parachutes. Two pipes emerging from the pressure vessel and protruding above the aerodisc had been used to cool the craft before separation. Lithium hydrate was carried internally to absorb some heat by changing into a liquid.

The lander came down on a 15–25° slope (measured by a "tiltmeter") at the base of a hill near Beta Regio. The photometer (measuring light levels) registered a cloud of dust thrown up by the touchdown. Surface pressure and temperature were 460°C and 90 atmospheres respectively. Windspeed was a gentle 1.4–2.5km/hr. The automatic timer ordered the single panoramic picture sequence to begin 15min later, apparently using a Luna 9/13-type scanning facsimile camera through a port under the disc once the protective cover had been ejected. The 5.8kg, 5W camera produced a 517-line black-and-white image with a 40° × 180° field of view, requiring about 30min to transmit. Black, white and grey rectangles on the lander ring provided a test pattern for calibration. The image relayed by the orbiter over a 1hr period (with breaks for other data) showed a rocky area with 30–40cm sharp stones and soil between them; the 10,000-lux floodlighting carried as a result of Venera 8's findings proved unnecessary. Dr Arnold Selivanov said that it was as bright as Moscow on a cloudy day in June. The Sun was 54° above the horizon at the time.

Gamma-ray spectrometers on both landers registered 0.3% potassium, 2×10^{-4}% thorium and 10^{-4}% uranium. Together with the density readings of 2.8 ± 0.1g/cc from the radiation densitometer, calculated from the scattering of radiation from a gamma source swung on to the surface by each craft, this indicated a basaltic terrain – at variance with the data from Venera 8, which suggested granite.

The capsule had been designed to survive for 30min, but a total of 53min of data was received and even then cutoff might have been caused only by the orbiter moving out of sight. The later Venera 11/12 flyby method permitted longer contact time. The Venera 9/10 buses continued their studies in orbit, and a Soviet announcement of March 22, 1976, declared that the main programme had ended but contact was still possible. The orbiters discovered a plasma and magnetic sheet draped like a hairpin over the planet, which has almost no intrinsic magnetic field; something similar would be found by ICE at Comet Giacobini–Zinner in 1985. The solar wind shockwave was only one-third of a planetary radius above the subsolar point. Direct measurements and radio occultations were used to study the ionosphere.

See Venera 10.

Venera 10

Launched: 0220 GMT June 14, 1975
Vehicle: Proton (D-1-e)
Site: Tyuratam
Spacecraft mass: 5,033kg at launch (1,560kg descent section, including 660kg lander; orbiter 2,314kg, plus 1,159kg propellant)
Destination: Venus
Mission: Orbiter/lander
Arrival: October 25, 1975 (landed 0517 at 16°N/291° longitude)

Top: *The first image from the surface of Venus. The lander ring is visible at bottom.*

Centre: *Venera 9/10 lander under ground tests.*

Bottom: *Venera 10's single panorama of its Venusian surroundings.*

Payload: Bus and lander as Venera 9
End of mission: October 25, 1975 (lander) and March 1976? (orbiter)
Notes: Second Venus orbiter and surface picture

Venera 10 followed its twin to Venus following ejection from a 162 × 206km, 51.52° Earth parking orbit, making course corrections on June 21 (14.5m/sec) and October 18 (9.7m/sec). The second burn targeted the descent section for entry over the planet's day side. Once it had been released five days later, with batteries charged and interior chilled, its bus executed a 242.2m/sec deflection manoeuvre to fly past the other side and brake with a 976.5m/sec burn into a 1,500km-perigee, two-day orbit. The path was later corrected to 1,400 × 114,000km, 49hr 23min and then 1,620 × 113,900km, 29.5° with the same period. On its way out to first apogee, the craft recorded the transmissions of the lander's descent for subsequent playback to Earth.

The sphere began entry at 0402 and the 23° path heated the protective shell to 12,000°C and produced a maximum deceleration of 168g. The exposed lander fell free once the parachutes were cut away at 49km (see Venera 9 for descent sequence) and its nephelometer/photometer found that the 30–40km-thick cloud layer ended at 30–35km. Temperature and pressure at 42km were recorded at 158°C and 3.3 Earth atmospheres, rising to 363°C and 37 atm 15km above the surface. On landing at 0517, some 2,200km from Venera 9's site (probes within each of the Venera pairs landed at similar longitudes becuse the planet rotates so slowly), the capsule found 465°C, 92 atm with a gentle 0.8–1.3m/sec wind. The single panorama, again taken without floodlighting, showed a weathered plateau with flat slabs of rock. Venera had come to rest on a 3m slab, listing several degrees backwards and to the left. There were no details visible on the horizon to the right, consistent with a gently rolling plain. It was clearly an older landscape than that surveyed by Venera 9.

A total of 65min of data was picked up by the orbiter, and again the lander might have actually survived for longer. See Venera 9 for spacecraft description and combined results.

Above: The Viking Lander. Lander 1 has been designated the "Thomas A. Mutch Memorial Station" in honour of the late leader of the lander imaging team. The National Air and Space Museum in Washington DC is entrusted with the safekeeping of the Mutch Station Plaque until it can be attached to Lander 1 by a manned expedition.

Right: Viking landing capsule system.

Viking 1

Launched: 2122 GMT August 20, 1975
Vehicle: Titan IIIE-Centaur D1
Site: ETR 41
Spacecraft mass: Orbiter: 2,328kg at launch
                      Lander: 663kg at launch, 612kg at landing
Destination: Mars
Mission: Orbiter/lander
Arrival: June 19, 1976 (landed July 20 at 22.483°N/47.94°W)
Payload: Orbiter:
    TV system
    Mars atmospheric water detector
    Infra-red thermal mapper
  Lander:
    TV system
    Gas chromatograph mass spectrometer
    X-ray fluorescence
    Seismometer
    Biological lab (pyrolitic release, labelled release, gas-exchange release)
    Weather station (temperature, pressure, wind velocity)
    Sampler arm
  Aeroshell:
    Retarding potential analyser
    Upper-atmosphere mass spectrometer
End of mission: Orbiter: August 7, 1980
                    Lander: November 13, 1982
Notes: First successful soft landing and surface images

The Viking Mars project emerged from the series of sophisticated Voyager planetary probes (not to be confused with today's Voyagers) that suffered cancellation by Congress in October 1967 on cost grounds. On December 4, 1968, the NASA Administrator approved a plan for two Viking 1973 missions to make high-resolution surveys of the planet and to send soft-landers to the surface in search of life. The orbiters would borrow from Mariner 9 technology but would have to be enlarged to carry the lander section. Martin Marietta received a $280 million contract on May 29, 1969, to build the landers, while the parent craft would be built at JPL; the whole project was directed by NASA's Langley Research Centre. The 1973 launch target slipped to summer 1975 as a result of general budget cuts, but Viking still ran out as the most expensive US planetary mission ever undertaken: $2,500 million at 1984 rates.

The Orbiter had to carry a much larger engine section than Mariner 9 because of its attached Lander. Two 140cm-long, 91cm-diameter tanks held a total of 1,600kg of monomethyl hydrazine and nitrogen tetroxide for the single 1,323N motor to provide a ΔV capability of 1,480m/sec. For attitude control there were two sets of six nitrogen jets at the solar panel tips, fed from two 30cm bottles.

Some 620W of power at Mars (1,400W at Earth) was generated by 34,800 cells covering a total of 15m² on four 123 × 157cm panels that gave Viking a giant span of 9.75m. Two 30A-hr rechargeable batteries were also carried. The Orbiter was completed by a typical Mariner octagonal body – 45.7cm deep and with sides alternating between 140cm and 51cm wide – housing 16 internal electronics bays cooled by thermal louvres. Spacecraft control was handled by two 4,096-word computers; dual tape recorders could each store 55 images for transmission to Earth at 4kbits/sec via the 20W, 147cm-diameter steerable S and X-band antenna. A low-gain S-band rod antenna could also be used, and a third mounted on a solar panel provided a lander link.

The 883kg (dry mass) Orbiter was very much oriented to the Lander's objectives, holding it dormant en route but continuously monitoring its health. The three instruments on the Mariner 9-type scan platform were intended to search for a smooth, warm and moist landing site to increase the chances of a safe touchdown in a region more likely to harbour life. The two identical 475mm-focal-length Cassegrain TV systems, using 38mm selenium vidicon tubes, could produce a resolution of 35m from a height of 1,500km. They would work alternately, one recording an image while the other took 4.48sec to read out 1,056 lines at 1,182 pixels/line, each image requiring 8.7Mbits.

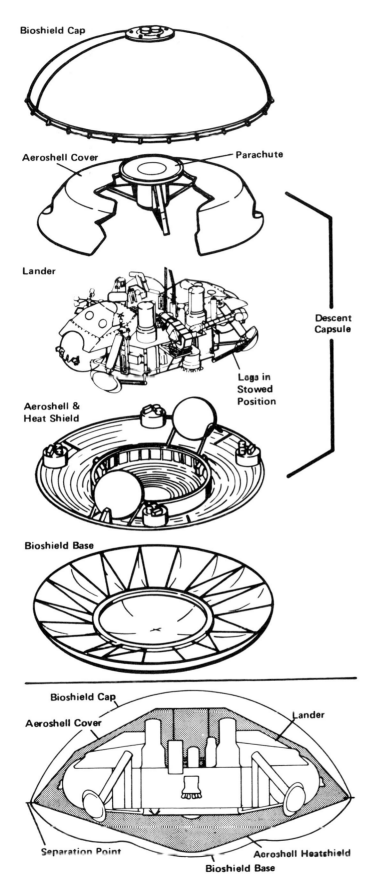

Bioshield Cap

Aeroshell Cover — Parachute

Lander

Legs in
Stowed
Position

Aeroshell &
Heat Shield

Descent
Capsule

Bioshield Base

Bioshield Cap

Aeroshell Cover — Lander

Separation Point — Aeroshell Heatshield

Bioshield Base

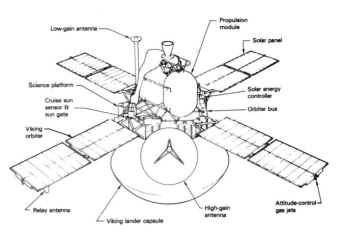

Low-gain antenna — Propulsion module — Solar panel
Science platform — Solar energy controller
Cruise sun sensor & sun gate — Orbiter bus
Viking orbiter
Relay antenna — Viking lander capsule — High-gain antenna — Attitude-control gas jets

Viking Orbiter and Lander in cruise mode.

Viking 1 was to have been launched on August 11, 1975, but problems with the launcher and then a malfunction that caused the spacecraft internal battery to drain to 9V from 37V forced engineers to swap over the "A" and "B" craft. Viking B (becoming Viking 1) was launched successfully into a 185km circular orbit and then injected into a Type II transfer path to Mars. The lens-shaped 3.66m fibreglass bioshield was ejected from around the Lander once the solar panels had deployed. The craft and its entry aeroshell had been sterilised to avoid Mars contamination, and the hermetically sealed biocap had to be carried within the Earth's atmosphere.

A 12sec course correction by the main engine at 1830 on August 27 yielded a Mars miss distance of 5,550km for June 19, 1976. Two additional corrections were made on June 10 (80m/sec) and 15 (60m/sec), mainly to bleed off dangerously high helium pressure in the main propulsion system. An initial correction of only 3m/sec had been required for trajectory adjustment. On June 19 Viking 1 fired its large engine for 38min to cut its approach speed of 14,400km/hr by 4,300km/hr and permit a 1,500 × 50,300km, 42.6hr orbit; 1,063kg of propellant was consumed. A trim manoeuvre on June 21 cut the orbit to 24hr 39min 36sec (to match the Martian day) and 1,500 × 32,800km, and the next day the first pictures of the A-1 landing site at 19.5°N/34°W were returned. Scientists had spent years deciding on this location in the Chryse region, so they were shocked to discover with Viking's better cameras that the surface was far from smooth. A July 4 Bicentennial landing was out of the question.

Using the new pictures, Viking's controllers at JPL decided on July 12 to make an attempt at 22.5°N/47.5°W on the western slopes of Chryse Planitia during July 20. The lander, encapsulated in its aeroshell, separated at 083214 that day and eight small thrusters fed from two hydrazine tanks fired at 083915 for 22min 48sec to send the combination into the atmosphere under control of the Lander's computer. The aluminium alloy shell protected the Lander from the 1,500°C heat of entry and also housed two scientific instruments of its own. The mass spectrometer analysed the atmosphere from 230km down to 100km and discovered nitrogen (2.5%) for the first time, a finding welcomed by biologists hoping to find signs of life. The temperature in the region averaged −93°C.

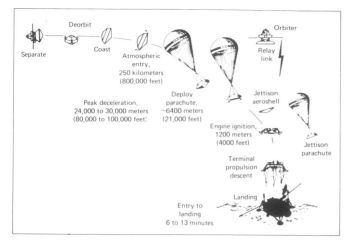

Diagram labels:

Separate — Deorbit — Coast — Atmospheric entry, 250 kilometers (800,000 feet) — Peak deceleration, 24,000 to 30,000 meters (80,000 to 100,000 feet) — Deploy parachute, ~6400 meters (21,000 feet) — Orbiter — Relay link — Jettison aeroshell — Engine ignition, 1200 meters (4000 feet) — Jettison parachute — Terminal propulsion descent — Landing — Entry to landing 6 to 13 minutes

Left: Viking landing sequence.

Below: The first picture from the surface of Mars, from Camera 2 of Lander 1 on July 20, 1976. The central rock is about 10cm across; at right is Footpad 2.

Bottom: The first picture returned by Lander 1's Camera 1 is this panorama of July 23. The horizon is 3km distant. The central rock is 3m across and 8m from the craft. The meteorology boom is deployed right of centre.

The maximum recorded deceleration was 8.4g at 27km. At 5.9km and 115105 the Lander's radar triggered the deployment of a 16.2m-diameter lightweight parachute on 30m lines. The radar ejected the aeroshell 7sec later and the Lander's legs deployed at 115124. By 1,463m speed was down to 54m/sec. At 1,400m, after parachute jettison, the three variable 276–2,840N engines cut in, drawing on 85kg of hydrazine; each had 18 small nozzles to minimise surface disturbance and contamination. Four 39N thrusters mounted on the hydrazine tanks themselves provided roll control. At 115306, 17sec late, Lander 1 made a safe touchdown at 2.4m/sec. But it was not until 121207 that telemetry back on Earth indicated the first successful Mars landing, just 28km off target.

The craft appeared much like a larger, updated Surveyor lunar lander. It was based on a six-sided aluminium-titanium box housing four of the experiments internally. The box was 46cm deep with alternating side lengths of 109cm and 56cm, the 1.3m legs and their 30.5cm-diameter footpads being attached to the short sides. It stood 2.1m tall and 3m wide, and massed 571.5kg empty. The entire Lander had been sterilised in an oven at 113°C for 40hr. A total of 70W of power for the wide range of instruments came from two plutonium-238 oxide radioisotope thermoelectric generators (RTGs) mounted on the top, and four nickel-cadmium batteries. The two sophisticated computers carried plated-wire memories of 18,400 words capacity each, sufficient to store commands for 60 days of activities. Some 198m of tape could hold 40Mbits of data for later relay to Earth using the 76.2cm steerable high-gain antenna working at 500 bits/sec and 2.2GHz. A direct link with Earth was possible for 2hr/day, and a fixed UHF transmitter permitted use of the Orbiter as a relay at 16kbits/sec. The upper deck carried a 2 × 2.5cm microdot bearing the names of 10,000 personnel associated with the project.

Seconds after touchdown, Camera 2 began a 5min scan in the area of Footpad 3. The two cameras, 80cm apart and 1.3m above the surface, were facsimile devices, each using a nodding mirror and rotating turret to reflect light on to a single photodiode for the scene to be built up strip by strip on Earth (as on the Soviet Lunas). Twelve sensors provided different capabilities: high-resolution black-and-white images; red, green and blue for full colour; and infra-red surface studies. Camera 2 showed that the pad had barely dented the dusty, rocky surface (Footpad 2 was completely submerged), and a 300° panorama later that day revealed a barren landscape, a desert strewn with rocks looking rather like a broken-up lava flow. The first colour signals came in the next day and were initially processed to yield a blue sky; in fact airborne dust particles produced a pink glow by scattering sunlight. The surface was also red, stained by rust (hydrated iron oxide). The pictures showed that the Lander was tilted 30°, with the cameras facing SE.

The instruments on the meteorology boom showed over the first few days a minimum temperature of −86°C before dawn and a maximum of −33°C in mid-afternoon, with a mean wind-speed of 29km/hr (gusts up to 51km/hr) and a pressure of 7.6mb. Pressure dropped by 30% in winter as carbon dioxide froze at the pole. Unfortunately, the internal seismometer, capable of registering a Level 3 quake 200km away, failed to unlock and could not operate. Its twin on Viking 2 was successful.

On July 22 the 3m robot arm was extended slightly to eject its protective cover, but a pin jammed and prevented full retraction. A second extension shook the troublesome item free. The first samples were collected on July 28 (Mars day 8) and deposited into experiment inlets on the Lander: one in the biology distributor, two into the gas chromatograph mass spectrometer (GCMS) and one into the X-ray fluorescence spectrometer. The latter irradiated its sample with X-rays from radioactive cadmium-109 and iron-55 sources and for 8hr recorded the X-rays emitted by the Martian material, revealing a composition of 2–7% aluminium, 15–30% silicon, 3–8% calcium and 12–16% iron.

The GCMS sample was ground up finely and heated to 200°C on August 6: little water was observed and no organic material. A heating to 500°C on August 12 produced copious amounts of water and some carbon dioxide but still no organics. A second sample behaved similarly. This was bad news for the biology team: how could there be life without organic material?

The other two samples were distributed equally between three experiments devoted to the search for life. The pyrolitic release experiment incubated its sample under a xenon lamp for five days with a carbon dioxide atmosphere labelled with radioactive carbon-14. If life was present, the carbon-14 would be taken into the soil sample and subsequently driven off by heating for detection. Seven of the nine samples taken by both Landers proved positive, but this was not conclusive: chemical processes cold have been at work.

The labelled-release experiment fed two drops of water and nutrients labelled with carbon-14 on to a sample for the first time on July 30. Bacteria should have consumed the nutrients and excreted detectable radioactive gases. Detectors did pick up a large initial count which then decayed over several days. The behaviour was untypical of a growing bacterial colony, and the suspicion is that a strong oxidising agent in the soil could have been responsible, especially when assimilation also took place in a sterilised sample.

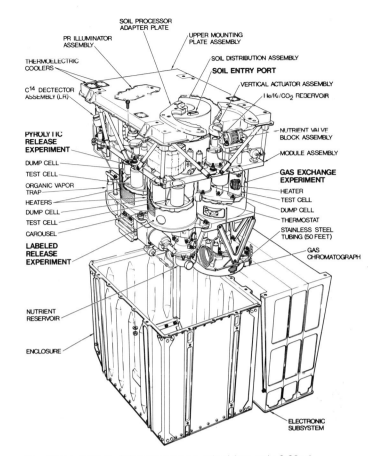

The Viking biology laboratory was packed into only 0.03m³.

The gas-exchange experiment applied unlabelled nutrient to a sample in a moist helium-krypton-carbon dioxide atmosphere, and signs of water, oxygen and methane were awaited. Oxygen was generated in profusion for a short period, as it was after sterilisation at 145°C, the conclusion being that some peculiar chemical reaction was occurring.

At a conference held in Washington DC during July 1986 to mark the 10th anniversary of the landings, biology team member Gilbert Levin reported that in the intervening decade no chemical reaction capable of explaining the labelled-release experiment results had been identified. He concluded that Viking did detect life after all, but other scientists remain unconvinced.

The Orbiter was shut down on August 7, 1980, having run out of attitude-control propellant, and the Lander was switched to making weekly weather reports until 1994 at least. After a transmission on November 13, 1982, however, engineers at JPL failed to regain contact in spite of persevering until May 1983. In all, both Landers returned more than 4,500 pictures – the last from Viking 1 on November 5, 1982 – and the Orbiters' 51,539 images mapped 97% of the surface at 300m resolution and 2% at 25m or better. A NASA Ames report of October 1985 concluded from an analysis of 20,000 orbital pictures that craters and other features above 30° latitude possessed softer outlines than those closer to the equator, possibly indicating sub-surface ice.

Above: *The 5,000km-long Valles Marineris dominates this Viking Orbiter 1 mosaic (north at top).*

Below: *Phobos imaged from 612km by Viking Orbiter 1 on October 19, 1978. Stickney, at 10km, is the largest crater on the moon and must have come close to disrupting the body. Soviet spacecraft are due to make a 1989 landing.*

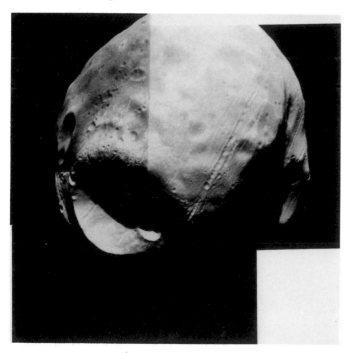

Viking 2

Launched: 1939 GMT September 9, 1975
Vehicle: Titan IIIE-Centaur D1
Site: ETR 41
Spacecraft mass: As Viking 1
Destination: Mars
Mission: Orbiter/lander
Arrival: August 7, 1976 (landed September 3 at 47.968°N/225.71°W)
Payload: As Viking 1
End of mission: Orbiter 0601 GMT July 25, 1978
Lander: April 12, 1980 (1,316 Earth days on surface)

The Viking A launch left the spacecraft with a Mars miss distance of 279,259km, but a course correction on September 19 allowed entry into orbit about the planet at 1329 on August 7, 1976, following a 39min 35sec, 1,100m/sec braking burn. As with Viking 1, the primary landing site at 44.3°N/10°W in Cydonia turned out to be too rough, and after a search Utopia Planitia at 47.89°N/225.86°W was chosen on August 30. It was hoped that a site more northerly than Viking 1's would have more moisture and therefore be more likely to harbour life.

At 201929 on September 3 Lander 2 and its aeroshell separated from the Orbiter and touched down at 223750 6,460km from Viking 1 on a rockier, flatter site, possibly among ejecta from the nearby 100km Mie impact crater. The surrounding rocks appeared alarmingly large and the Lander was tilted at 8½° to the west. JPL had lost contact just 26sec after separation when the Orbiter antenna broke Earth lock, and communications were not regained until shortly after the automatic landing.

Acquiring a sample in the area dubbed "Bonneville Salt Flats" by scientists. The foreground rock is 20cm high. The resulting trench is about 8cm wide.

The biology experiments produced much the same results as those of Viking 1. The seismometer, uncaged successfully this time, detected a 2.8 Richter quake 110km away, the signal's rapid decay suggesting water or frost in the 14–18km-thick crust. It was also sensitive enough to detect winds and the tape recorders' rumblings. The meteorology instruments recorded night-time temperatures of −87°C early in the mission. Later, in winter, a low of −118°C was reached, and frost was seen on the surface for the first time in mid-September 1977. Surface pressure ranged from 7.4 to 10mb.

The Orbiter approached to within 28km of the Martian moon Deimos in May 1977 and found a surface with many dust-filled, subdued craters. Deflection of the trajectory indicated a density of less than 2g/cc, typical of a carbonaceous chondrite asteroid.

The two Orbiters showed Mars' residual north polar cap to have a temperature of around 200 K. This is too warm to be frozen carbon dioxide, so the cap must therefore be water ice. The Lander detected krypton and xenon in the atmosphere for the first time, and combined results show a surface atmosphere of 95.32% carbon dioxide, 2.7% nitrogen and 1.6% argon. Water vapour is 0.03%.

The Orbiter lost most of its remaining attitude-control gas in a series of leaks during February and March 1978, and operations ceased that July. The Lander continued relaying weather reports until it was shut down on April 12, 1980, after 1,316 Earth days on the surface. See Viking 1 for description of spacecraft and instruments.

Luna-1975A Mass 5700kg? Launched October 16, 1975, by D-1-e from Tyuratam but failed to reach Earth orbit. Probably second attempt at Luna 23 sample return objective.

The second picture taken by Viking Lander 2 was this 330° panorama running from north-west at left through to south-east at right. The craft's 8° tilt to the west produced the undulating horizon. At left is the cover for the sampler arm.

Luna 24 marked the end of initial scientific exploration of the Moon for both the USSR and US. Its predecessor's failure to obtain a sample from the previously untouched Mare Crisium was clearly considered to leave such an important gap in lunar knowledge that a back-up spacecraft had to be flown (see also Luna-1975A). Why there should have been such a lengthy delay is difficult to explain, although there was a general slowing down of planetary launches in the 1970s. NASA saw its own launch rate fall drastically from the heady heights of the 1960s and early 1970s because of financial limitations. The Soviets could have suffered a similar problem, having to work a back-up mission into a tight budget. Luna 24 remains the only Soviet probe not to be targeted at Venus since October 1974; the next lunar flight is not expected until the early 1990s.

A 188 × 243km, 51.54° Earth parking orbit was followed by the Proton injection burn and a course correction by the spacecraft itself on August 11. At 2311 on August 13 a braking burn established a circular 115km lunar orbit with a period of 119min and inclined at 120° to the equator. A 12 × 120km path resulted from manoeuvres on August 16 and 17 in preparation for the descent burn to begin at 0630 on August 18 and a successful landing at 0636 near the Luna 23 site in the SE region of Mare Crisium. This time the 2m rotary percussion drill was not damaged and on command from Earth it extracted a full core. The Luna 16 and 20 drills had dropped

The sample returned by Luna 24.

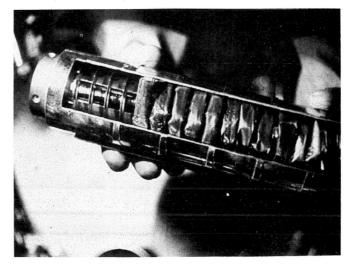

Luna 24

Launched: 1504 GMT August 9, 1976
Vehicle: Proton (D-1-e)
Site: Tyuratam
Spacecraft mass: 5,700kg? at launch (1,880kg? on surface)
Destination: Moon
Mission: Sample return
Arrival: Landed August 18, 12.75°N/62.20°E; Earth return August 23
Payload: As Luna 23
End of mission: August 23, 1976, capsule landed 0555 200km SE of Surgut in Western Siberia
Notes: Eighth Soviet lunar lander, third successful Soviet sample return. Remains final lunar spacecraft

significant proportions of their samples, so Lunas 23 and 24 incorporated a new design that also permitted a much deeper penetration. As the drill worked, the soil was pushed into an 8mm-diameter, 1.6m-long flexible tube which was then coiled like a spring for storage in the capsule.

Some 170g had been loaded aboard when the ascent stage ignited at 0525 GMT on August 19 to boost the cargo to a speed of about 2.7km/sec for the four-day uncorrected flight to Earth. Contact was maintained with the descent stage in order to continue testing the drill. The separated capsule hit the atmosphere at a steep angle, decelerating rapidly and throwing out its drogue parachute at 15km altitude and main 'chute at 11km for a safe landing. The ninth and so far final lunar sample (six Apollo, three Luna) was taken to the Vernadsky Institute of Geochemistry and Analytical Chemistry in Moscow, where a quick inspection showed it to be silvery/grey with a brownish tint, similar to that returned by Luna 16 but more powdery and with a more extensive sprinkling of large grains. It appeared to be layered, as if it had been laid down in successive deposits. NASA's Dr Michael Duke returned from Moscow in December 1976 with two ½g samples given in exchange for Apollo material. The three unmanned landers returned a total of about 320g, compared with 384kg from the six Apollos.

Voyager 2

Launched: 142945 GMT August 20, 1977
Vehicle: Titan IIIE-Centaur D1 + TE 364-4
Site: ETR 41
Spacecraft mass: 825kg at launch
Destination: Jupiter, Saturn, Uranus, Neptune
Mission: Flyby
Arrival: Jupiter: July 9, 1979, closest approach 645,000km, 2229 GMT
Saturn: August 26, 1981, closest approach 101,000km, 0121 GMT
Uranus: January 24, 1986, closest approach 71,000km, 1759 GMT
Neptune: August 25, 1989, closest approach 44,800km (planned)
Payload: As Voyager 1
End of mission: 2013?
Notes: First Uranus and (projected) Neptune flybys, discovered 14th Jovian satellite, third Jovian ring component, 10 Uranian moons, two Uranian rings

Voyager 2 was actually the first of the outer-planet pair to be launched, but its speedier twin overtook it in the Asteroid Belt on December 15, 1977 (see Voyager 1 for spacecraft description and project history). All of the instruments had been switched on and checked by September 2. A potentially disastrous event occurred on April 5, 1978, when the primary radio receiver aboard failed and the Central Computer and Sequencer had to switch automatically to the back-up unit, which has been in use ever since. Should that drop out of commission, the spacecraft cannot be fully programmed for the final encounter with Neptune in 1989. In addition, a fault limits its bandwidth to only ±96Hz instead of the design value of 1,000Hz.

Beginning on April 24, 1979, Voyager 2 relayed Jovian

images for time-lapse movies of atmospheric circulation, and as it closed in it was apparent that the planet had changed appearance in the four months since Voyager 1's visit. The Great Red Spot had become more uniform, for example. The previous encounter had prompted two major revisions to the flightplan: a 10hr volcano watch on the moon Io and a more detailed study of the ring. Voyager 2 approached no closer than a million km to Io but the face had clearly altered and six of the earlier plumes were still active, although Pele appeared dormant.

The closest approach to Callisto (215,000km) was made at 1313 on July 8, and the view of the previously unseen hemisphere revealed a densely cratered surface, unchanged for possibly thousands of millions of years. One picture taken that day was studied in detail in early October, revealing Jupiter's

Below: *Voyager 2 will probably remain the only mission to Uranus this century. Its close encounter with Miranda revealed a chaotic surface exhibiting a diverse range of features.*

Bottom: *Jupiter's ring, illuminated by forward-scattered sunlight, was seen particularly well by Voyager 2.*

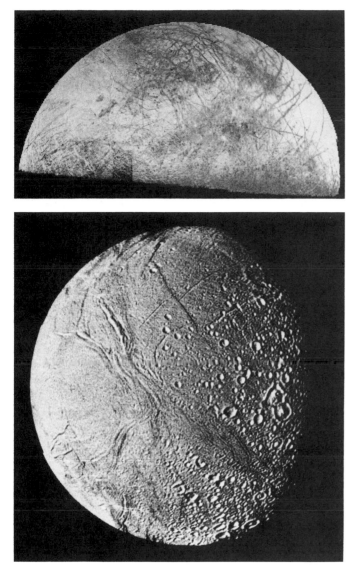

that reprocessing of one of the outward-bound images had revealed a faint third component to Jupiter's rings, stretching out to 210,000km. Satellites Amalthea and Thebe possibly contribute some of the material.

Firing the thrusters for 76min only 2hr following Jovian periapsis, rather than two weeks later as normal, put Voyager on target for Saturn and saved 10kg of propellant for Uranus.

The craft began its encounter with Saturn on August 22, 1981, when it imaged the moon Iapetus from 900,000km. But the distance was too great to explain the large dark-brown stain on the leading hemisphere. Hyperion appeared as a battered 300km-diameter lump that looked as if it had suffered a severe collision. Voyager 1's results prompted a thorough study of the ring system, and for 2hr 20min around closest approach the photopolarimeter was used to observe the star Delta Scorpii through the rings, measuring the flickering light level. The 100m resolution was ten times better than was possible with the cameras, and many more ringlets were discovered. The A-ring turned out to be no thicker than 300m.

Enceladus was revealed to be similar to Ganymede, its cratered areas contrasting with smooth and icy regions. Tethys was found to possess a 400km crater but the best images were lost when the azimuth control motor of the scan platform stuck during Earth occultation behind Saturn. Subsequent analysis suggested that extensive use of the high-rate, 1°/sec slewing mode had jammed the gearing mechanism as the lubricant lost its effectiveness. By the time the platform was working again two days later in a slower mode a lot of valuable data had been lost.

The final major event was the 2.2 million km flyby of Phoebe on September 4: the distant 200km-diameter body appeared as little more than a collection of dots.

Voyager 2 broke new ground over the next 4½ years as it headed towards distant Uranus at the end of a 33 AU journey. Little was known of the planet – not even its rotational period – nor of its five moons and nine very dark, thin rings. Voyager 2

Top: *Voyager 2 showed Europa's ice-covered surface to be riven with an intricate network of cracks.*

Above: *Enceladus recorded by Voyager 2 from 119,000km.*

Right: *The 1986U7 and U8 moons shepherding Uranus' bright Epsilon ring were discovered in this January 21, 1986 image, recorded at a distance of 4.1 million km and a resolution of 35km. Inside are the Delta, Gamma, Eta, Beta and Alpha rings. The faint 4, 5, and 6 rings are barely visible on the original.*

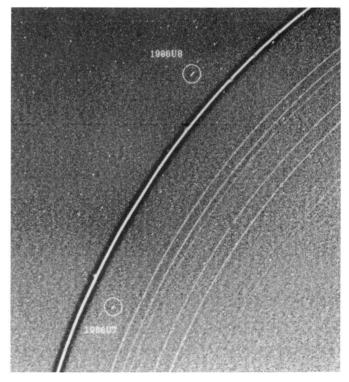

14th satellite. Now officially named Adrastea, it is only 30–40km in diameter and orbits close to the ring.

Ganymede passed within 62,000km at 0806 on July 9, and Voyager discovered a large dark region since named Regio Galileo. Europa followed at 1843 and a distance of 206,000km, sufficiently close for the streaks seen by Voyager 1 to be resolved into a myriad of cracks in a thick covering of ice. No relief of any kind was evident: the moon was "as smooth as a billiard ball", according to one scientist. (Arthur C. Clarke, in his novel *2010*, had Europa supporting life in the warm waters of its oceans below the ice.) It was announced in August 1985

would fly through the system just inside the moon Miranda's orbit like an arrow speeding at a target. There would be only 5½hr of close encounters in which to gather data, and only the moons' southern hemispheres would present themselves for scrutiny. This path was determined by Neptune mission requirements, and the "observatory" phase of long-distance studies began on November 4, 1985, as the gap closed at 15km/sec. Signals at this stage took 2hr 25min to reach Earth, 2,880 million km away, and the great distance permitted a data rate of only 21.6kbits/sec. Light levels were 400 times less than under terrestrial conditions, and the whole spacecraft had to be moved bodily to avoid image smearing at closest approaches (the scan platform was treated carefully and not used at high slew rates).

The first major result was the discovery of 10 new moons, the largest (1985U1, unofficially named Puck) only 170km across, between December 30, 1985, and January 23, 1986, all inside the orbits of the five known satellites. The eighth and ninth flank the Epsilon ring, acting as shepherds, and the rings themselves appear to be composed of mainly boulder-sized particles. The Eta ring disappears in places. Two further rings were identified by observing occultations of the stars σ Sag and Algol, and the plasma wave instrument picked up 20–30 dust impacts as Voyager 2 crossed the ring plane. The rings are deficient in small particles but the impacts indicated a 4,000km-thick band of μm-size material with a density of about one per 1,000m³. A 96sec time exposure taken from behind the planet showed a complex web of dust structures backlit by the Sun. The known moons are a disparate bunch, with Miranda the strangest of all. It is covered in faults, grooves, terraces and a 16km-high cliff, and seems to have examples of all the geology found elsewhere in the solar system. Geologists consider it to have fragmented at least a dozen times and reformed in its present confused state.

Oberon is heavily cratered and possesses a 6km mountain – an unusual feature for a moon – and cratered and fractured Ariel is the brightest of all (again, it appears to have fragmented and reformed). Titania appears to have frost-like material oozing out of faults and is covered in bright 10–50km

craters possibly created by impacting comets. Umbriel is the odd one out. Its dark and severely pockmarked surface suggests it is the oldest of the major Uranian moons.

Uranus itself displayed little detail beneath its 85% hydrogen/15% helium atmosphere (26% helium by mass). Curiously, the 60K average temperature is the same at the Sun-facing south pole as at the equator. There is also a haze layer around the pole and Voyager 2's detection of a magnetic field revealed the internal rotation period to be 17.24hr. Tracking atmospheric features showed a differential rotation, from 16.3hr at 35° latitude to 14.2hr at 70°. Winds blow at 15–220m/sec in the same sense as Uranus turns.

The magnetic axis (the field is stronger than Saturn's) is unusually tilted 55° below the spin axis, so that the 10 million km-long magnetotail is twisted into a helix away from the Sun. Uranus also produces profuse UV emission on the daylight side, as seen at Jupiter and Saturn.

Voyager 2 remains in good health and looks set to reach Neptune in working order on August 25, 1989. A 2½hr, 21.1m/sec burn expending about 12kg of hydrazine on February 14, 1986, adjusted its path to allow a 1,300km flyby of Neptune and one of 6,000km at Triton, the closest of all its many encounters. Mission planners subsequently decided to move the flyby to 4,700km above the cloudtops, partly to avoid a suspected ring system, yielding a 40,000km encounter at Triton. A Backup Mission Load (BML) of about 1,000 words was loaded into the computers in September 1986 to execute an automatic survey should the remaining radio receiver fail and prevent full contact three years later.

Beyond Neptune, Voyager 2 will join its twin and the two Pioneers in investigating the heliosphere, the final encounters over the planet's north pole deflecting it 48° below the ecliptic. Power assessments suggest that its signals will be detectable until 2013 at a distance of 106 AU, its next close approach to a star being a 0.8-light-year flyby of Sirius in 358,000 years' time.

Voyager 1

Launched: 125601 GMT September 5, 1977
Vehicle: Titan IIIE-Centaur D1 + TE 364-4 kick stage
Site: ETR 41
Spacecraft mass: 825kg at launch
Destination: Jupiter, Saturn
Mission: Flyby
Arrival: Jupiter: March 5, 1979, closest approach 280,000km at 1205 GMT
Saturn: November 12, 1980, closest approach 124,000km at 2345 GMT
Payload: Imaging system (38.2kg)
Infra-red spectrometer (IRIS) (30.2kg)
Ultra-violet spectrometer (UVS) (4.5kg)
Photopolarimeter (4.4kg)
Planetary radio astronomy (7.7kg)
Magnetometers (5.6kg)
Plasma particles (9.9kg)
Low-energy charged particles (7.5kg)
Plasma waves (1.4kg)
Cosmic ray telescope (7.5kg)
End of mission: 2012?
Notes: Discovered three Saturnian moons, 2 Jovian moons, Jupiter's ring, Io's volcanoes and fine structures of Saturn's rings, probed Titan's atmosphere

Voyager 1's encounter with Saturn in 1980 pushed it north of the ecliptic, while Voyager 2 was precisely targeted to hold it within the ecliptic for the Uranus and Neptune legs.

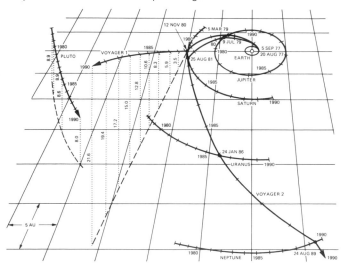

The dual Voyager missions to the outer solar system have generated probably the most exciting results of the entire deep-space programme. The quantity of data from five planetary encounters, with a sixth to follow in 1989, is certainly unprecedented.

Montage of Voyager 1 Jupiter images. At left is Io, bottom Ganymede, centre Europa and right Callisto. The Great Red Spot and complex circulation patterns are evident in Jupiter's atmosphere.

The "Mariner Jupiter Saturn 1977" mission emerged from studies in the 1960s revealing that unusual planetary alignments during 1976–80, occurring only once every 176 years, would allow probes to swing by several planets while reducing journey times to reasonable levels for the three outer planets. The opportunity was too good to pass up (the Soviets do not appear to have considered the option) and the "Grand Tour" was born, with a 1977 launch leading to Jupiter (1979), Saturn (1980) and Pluto (1986), and a 1979 start producing Jupiter (1981), Uranus (1985) and Neptune (1988).

But the mission grew ever more complex and costs escalated towards the $1,000 million mark (1970 values) until budget cutbacks forced its cancellation on January 24, 1972. NASA/JPL immediately concentrated on producing a mission at one-third the cost, drawing on the Mariner and Viking designs and focusing on Jupiter, Saturn, moon Titan and Saturn's rings. This reduced effort was approved by NASA in May 1972 shortly before the Pioneers blazed the trail to the gas giants. The missions would end formally at Saturn but the

scientists still had their collective eye on Uranus and Neptune encounters if the spacecraft and funds held out. The "Voyager" tag was adopted in 1977, resurrecting an old Venus-Mars probe name.

Voyager's appearance is dominated by a 3.66m-diameter aluminium honeycomb X and S band dish and the three booms protruding from the 10-bay body below it. The 23W X-band (8.4GHz) transmitter could return data at 115.2kbits/sec at Jupiter (less at more distant targets), with a 500Mbit tape recorder able to store upwards of 100 images. The dish antenna is fixed so that the Voyagers spend most of their time pointing at Earth except during special manoeuvres such as course corrections.

Most of the electronics are carried in the ten 47cm-high bays (Bay 1, for example, has the radio transmitter) in which they were protected against Jupiter's harsh radiation environment. The thermal blanketing also carries several layers of micro-meteoroid protection.

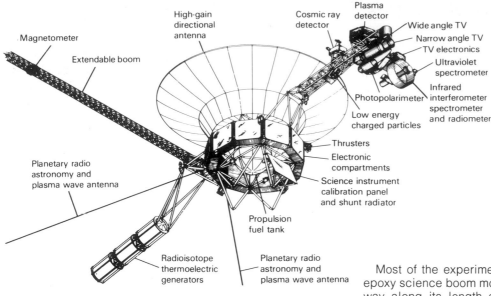

Magnetometer

Extendable boom

High-gain directional antenna

Cosmic ray detector

Plasma detector

Wide angle TV

Narrow angle TV

TV electronics

Ultraviolet spectrometer

Infrared interferometer spectrometer and radiometer

Photopolarimeter

Low energy charged particles

Thrusters

Electronic compartments

Science instrument calibration panel and shunt radiator

Planetary radio astronomy and plasma wave antenna

Propulsion fuel tank

Radioisotope thermoelectric generators

Planetary radio astronomy and plasma wave antenna

Three computers control the spacecraft. The Computer Command System (CCS) issues orders to the others and acts as a fault detector. It carries two redundant memories with 4,096 18-bit words each, half taken up with fixed instructions (e.g., for decoding commands from Earth) and the rest available for reprogramming throughout the mission. The 98-day encounter period at Jupiter, for example, required 18 loads of 1,290-word command sequences from Earth, though cruise periods are less demanding. The Attitude and Articulation Control Subsystem (AACA) looks after the pointing of the scan platform and the spacecraft as a whole, while the Flight Data Subsystem (FDS) collects the science and engineering data and formats them for transmission.

Attitude control depends on Sun and Canopus sensors, with three gyros available for inertial reference. There are 16 0.89N thrusters drawing on 104kg of hydrazine from a 71cm spherical titanium tank nestling in the centre of the 188cm-diameter main body to provide a total ΔV capability of 190m/sec. At launch the whole vehicle weighed 2,066kg, with an integrated solid-propellant kick stage mounted on struts below the aluminium body. Firing for 43sec, the motor used four 445N thrusters for thrust-vector control and four 22N thrusters for roll control (drawing on the spacecraft supply), and was then discarded.

Power is generated by three 46cm-diameter plutonium oxide RTGs mounted at the end of a boom (to minimise irradiation of the electronics and instruments), decreasing at 7W/year from the 450W (converted from 7kW of heat) available at launch.

Each Voyager also carries a sample of terrestrial culture should it ever be retrieved by an alien race in the distant future. A 30cm gold-plated copper disc, together with a needle and playing instructions, is mounted on the body casing. On it are recorded natural Earth sounds, 90min of music, 115 analogue pictures and greetings in 60 languages. The aluminium cover should provide protection against occasional dust impacts for 100 million years.

Most of the experiments are attached to a 2.3m graphite epoxy science boom mounted 180° from the RTG boom. Midway along its length are the cosmic ray and low-energy charged particle detectors, with the plasma science instrument further out. At the end is the two-axis scan platform housing the TV system, photopolarimeter, IRIS and UVS. The close approach to planets and moons requires platform slewing during exposures to minimise image smearing. The plasma wave and planetary radio astronomy devices share two 10m beryllium-copper aerials mounted below the dish at 90° to each other. Four magnetometers reside on their own 13m, 2.26kg epoxy glass boom.

Voyager 1 was actually launched after Voyager 2 because it would follow a slightly faster route and overtake its twin in the Asteroid Belt. The Centaur accelerated the spacecraft from parking orbit, and 15sec after separation the 68,050N kick stage ignited to exceed Earth escape speed. Eleven minutes later the propulsion module was ejected and Voyager sailed on to cross the Moon's orbit in only 10hr.

Imaging of Jupiter began in April 1978 from 265 million km out, and by January 6, 1979, pictures were being relayed back every 2hr to produce time-lapse motion studies of the rotating, swirling atmosphere. It was already clear that Jupiter was more active than during the Pioneer passages, and that the earlier atmospheric models were inadequate to explain the new features.

By February the looming planet had grown too large to fit in a single TV frame, and on February 10 Voyager 1 crossed the orbit of outermost moon Sinope, travelling at 13km/sec and still some 23 million km from closest approach. On February 28 the instruments revealed that the spacecraft had passed from interplanetary space to Jupiter's magnetosphere (10 times the size of the Sun), although the solar wind pushed the planet's magnetic field boundary to and fro several times.

The first major discovery came a day before closest approach, when an 11.2min exposure with the 1,500mm camera displayed a fine ring of material with an interior halo orbiting the planet. It was no more than 30km thick and extended from an altitude of 57,000km down to cloud level (Voyager 2 discovered a third component). Jupiter now joined Saturn and Uranus as a ringed planet, strengthening speculation that Neptune will soon join the club.

Voyager accelerated in past Amalthea and revealed a small, irregular, reddish lump of rock. The next moon to be

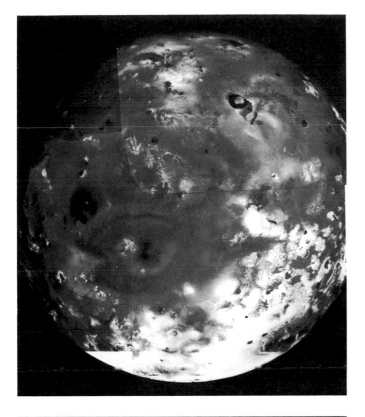

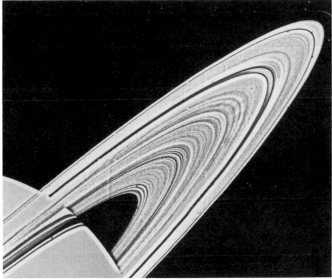

Left: Mosaic of Io images taken from 376,950km on March 5, 1979, revealed a young, active surface.

Below left: The incredibly complex Saturnian ring structure is highlighted in this two-image mosaic, produced on November 6, 1980, from 3 million km. The faint F-ring can be seen at the tip, together with the planet's 14th satellite, discovered by Voyager 1.

The 21,000km approach to Io was followed by occultation behind Jupiter's limb, allowing S and X-band transmissions to probe the atmosphere. Aurorae and lightning flashes could be seen lighting up the night side. While it was still in darkness, Europa hove into view 750,000km away and presented a light surface covered in streaks. Voyager 2's closer visit exposed an intricate network of cracks in a frozen ocean.

A 112,030km approach to Ganymede, at 5,270km diameter the largest moon in the solar system, followed at 0235 on March 6. Its grooved and cratered surface was not unlike the Moon's, but it was Callisto that provided the surprise. The 126,000km encounter at 1750 GMT on March 6 revealed a huge bullseye of an impact basin some 1,500km across (since named Valhalla).

Callisto was the final port of call before Voyager 1 sped away, making observations over its shoulder as it went. Two new moons, 1979J2 and J3, were discovered while data were being checked through during 1980, bringing the known total to 16. They have since been named Thebe (80km diameter) and Metis (40km), respectively. A radiation dosage one thousand times the lethal level for humans had caused electronics damage, resulting in serious degradation of some of the highest-resolution images of Io and Ganymede.

The encounter increased Voyager 1's speed to 84,500km/hr and a course correction on April 9 set the probe up for a flyby of Saturn 800 million km further on. A 13.7min, 2m/sec burn was carried out on October 10 to avoid impact with Titan.

The Saturn meeting in November 1980 produced results no less fascinating. The ring system was found to consist of thousands of bands. The F-ring seemed to be three intertwined ringlets, and a G-ring was confirmed. Particles ranged in size from 10m water icebergs down to 0.0005cm specks. "Spokes" – explained as short-lived micron-sized particles levitated electrostatically above the ring plane – appeared to rotate with the B-ring. Shepherding satellites Prometheus and Pandora were discovered just either side of the F-ring, and a third, 30km-diameter Atlas, was found on the sharp outer A-ring boundary. These satellites clearly help to keep the rings well defined. The F-ring is now believed to be suffering impulses from unseen moons in eccentric orbits, and there are a dozen gaps throughout the ring system that could harbour moons awaiting discovery by future spacecraft.

Voyager 1 observed all of Saturn's known moons except Phoebe from varying distances: Mimas 88,000km, Enceladus 202,000km, Tethys 400,000km, Dione 161,000km, Rhea 72,000km. Enceladus appeared smooth (Voyager 2 passed closer and found craters) but the others were well cratered and all appeared to be largely water ice. Enceladus has the brightest known surface in the solar system and could also be the source of material for the E-ring. Mimas, only 390km in diameter, has a 130km-diameter peaked crater (Herschel) dominating one hemisphere. Tethys has a great 750km fracture, named Ithaca, and Dione and Rhea both have wispy markings that could be ice migrating from their interiors.

imaged was Io, expected to present a cratered, lunar-like face to the cameras but emerging instead as an iridescent sulphur-covered world of yellows, oranges and browns with at least eight active volcanoes spewing material into space (the first eruption was seen in an exposure taken on March 8 for navigational purposes). It is probably the most geologically active body in the solar system, a massive amount of internal heat being generated by a 75m distortion in its diameter every 42hr as it orbits Jupiter elliptically. The volcanoes generate a torus of sulphur and oxygen ions at temperatures of 100,000°C around the moon's orbit, where they are accelerated up to 10% of light speed to produce aurorae in Jupiter's atmosphere.

Montage of the Voyager 1 Saturn encounter. At foreground is Dione, with Tethys and Mimas to right and Enceladus and Rhea to left. At top right is Titan.

Titan was the major target among the moons. The close approach, 4,000km at 0541 on November 12, showed it to be covered in an optically thick atmosphere completely obscuring the surface, with a dark "hood" encircling the north pole. Analysis of radio occultation data showed the 90%-nitrogen atmosphere (not methane, as expected) to have a surface pressure 1.6 times that on Earth and a surface temperature of −180°C. There are possibly methane clouds 10–15km above the surface; argon is the third major constituent. The diameter of 5,140km makes it slightly smaller than Ganymede and there was no measurable magnetic field. There is an enormous cloud of uncharged hydrogen surrounding Titan and its orbit, extending inwards to Rhea's path.

Voyager 1's scan-platform instruments were turned off on December 12, 1980, but the fields and particles equipment continues to take data as the craft leaves the solar system. The requirements for a close Titan flyby had prevented a retargeting to Uranus and Neptune. Departing 35.5° north of the ecliptic, Voyager 1 is heading towards the boundary at which the Sun's magnetic envelope is pushed back by the interstellar wind. But it is debatable whether it will reach this region in working order: signal power should drop below the detectability level in 2012 at a distance of 121 AU. Tracking of the four Pioneers and Voyagers could help to locate a 10th planet or solar dark star companion by measuring differential deflections of their trajectories. If Voyager 1 were visible, it would appear to be moving slowly northwards between the constellations of Ophiuchus and Hercules, heading for a 3-light-year encounter with the ageing M-type dwarf star AC+79 3888 in Camelopardus in 40,000 years' time.

See Voyager 2.

Pioneer Venus 1 (Pioneer Venus Orbiter)

Launched: 1313 GMT May 20, 1978
Vehicle: Atlas-Centaur 50 (Atlas No 5030D)
Site: ETR 36A
Spacecraft mass: 553kg at launch
Destination: Venus
Mission: Orbiter
Arrival at target: December 4, 1978
Payload: Cloud photopolarimeter (5kg)
Surface radar mapper (9.66kg)
Infra-red radiometer (5.9kg)
Airglow ultra-violet spectrometer (3.08kg)

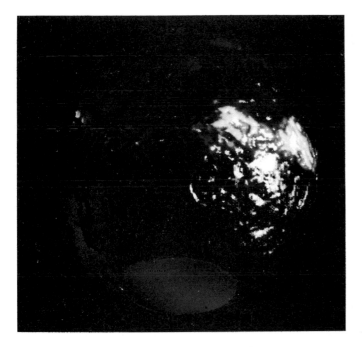

Neutral mass spectrometer (3.81kg)
Solar wind plasma analyser (3.90kg)
Magnetometer (2.01kg)
Electric field detector (0.80kg)
Electron temperature probe (2.16kg)
Ion mass spectrometer (2.99kg)
Charged-particle retarding-potential analyser (2.81kg)
Gamma-ray burst detector (2.80kg; non-Venus experiment)
End of mission: 1992?
Notes: Third Venus orbiter (first US), first surface radar mapper

The NASA Ames Research Centre's Pioneer Venus programme was designed to sample the Venusian atmosphere directly and study the surface (by radar) and clouds in the long term. An early-1970s proposal suggested that two carriers, each with four atmospheric probes, should be launched during the December 1976–January 1977 window, followed by an orbiter in May–August 1978. The project evolved into one multiple carrier and an orbiter, both timed for the 1978 opportunity and using Atlas-Centaur instead of the less powerful Delta. Congress approved the project in August 1974 and Hughes was appointed main contractor the following November.

STAR SENSOR, TANKS, AXIAL THRUSTERS, AND AFT OMNI ANTENNA SHOWN ROTATED FOR CLARITY

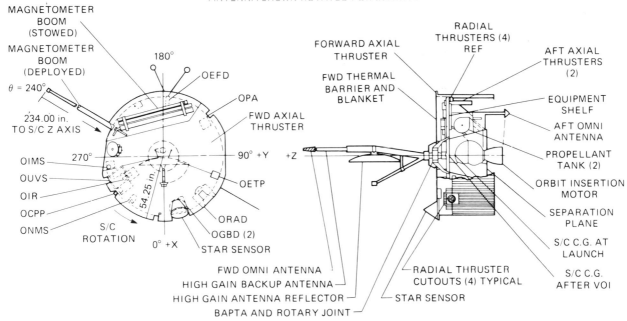

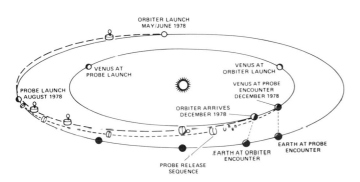

Top left: *PVO's modest radar system permitted the first extensive surface map to be generated. The blank area around the south pole represents incomplete coverage. The high Aphrodite region is visible at left and centre.*

Above: *Pioneer Venus Orbiter (PVO) principal features.*

Left: *The Pioneer Venus trajectories.*

Both spacecraft adopted a basic 2.53m-diameter, 1.22m-high cylindrical bus, spin-stabilised (as were all Pioneer-class craft) at 5rpm at Venus and 0.5rpm during cruise. The orbiter bus centres on a conical thrust tube, the base of which interfaced with the launcher. At its centre is a 17,800N Thiokol solid-propellant retromotor, designed to brake the craft by 3,780km/hr into a 24hr Venus orbit. Attitude control is maintained by seven hydrazine thrusters (three axial, four radial) drawing on 32kg of propellant in two 32.5cm tanks. The thrusters maintain spin and change attitude and orbit using a star sensor and Sun sensors for reference.

An equipment shelf near the top is supported by 24 struts attached to the thrust tube, holding the electronics and 45kg of scientific instruments. Two data storage units provide a 1,048,576-bit capacity, sufficient to hold data generated during Earth occultation by Venus (which can last up to 26min). Up to 312W of power was provided by 7.4m² of solar cells (14,580 of them) on the outer drum, with two nickel-cadmium 7.5A-hr batteries providing eclipse power. The 1.09m-diameter despun S and X-band high-gain parabolic antenna is mounted on a 2.99m-long central boom that also carries a back-up sleeve dipole antenna and an omni antenna (a second is carried under the bus). The 750mW X-band transmitter was used solely to probe the atmosphere during radio occultation as viewed from Earth. All of the experiments are mounted on the equipment shelf, with the magnetometer using a 4.72m deployable radial boom, and are thus scanned across the planet as the orbiter rotates.

PVO was launched on a Type II trajectory to yield a lower arrival speed: a faster Type I path would have required a retromotor accounting for 50% of the total mass. At 1551 GMT

Full-disc Venus image acquired on March 3, 1979, from 66,000km.

December 4, 1978, 56 million km from Earth, PVO began a 23min occultation behind Venus and automatically fired the retromotor for 30sec at 1558, emerging into Earth view again at 1614. Tracking showed the initial orbit to have a perigee of 378.6km (planned 350.0km), an inclination of 105.02° and a period of 23.4hr instead of the planned 24hr. Two revolutions later the hydrazine thrusters were fired to produce a 24hr period so that perigee/apogee occurred at the same time each Earth day.

Over the first 16 orbits, seven burns reduced perigee height to 150km (apogee 66,889km). The instruments had been switched on around the first apogee and the first image from the cloud photopolarimeter was received on December 6. Radar sweeps were taken only around perigee to produce crosstrack resolution of 6.9km, and the stored data replayed to Earth on the way to apogee. The UV cloud images were generated towards apogee, each taking about 3½hr to build up. 350 UV images were returned during the primary mission of December 1978–August 1979.

The radar mapper, which malfunctioned from December 18, 1978–January 20, 1979, allowed a topographical map of most of the surface between 73°N and 63°S at a resolution of 75km to be constructed. It showed Venus to be smoother and more spherical than Earth: 60% of the surface is within 0.5km of the mean surface. There is no polar flattening or equatorial bulge, consistent with the slow 243-Earth-day rotation rate. Earth-type features such as continents, mountains and rift valleys indicate tectonic activity. Maxwell Montes, at 10.8km, is the highest point on the planet; Diana Chasma is 2.9km below the mean surface. Beta Regio has two large shield volcanoes, now named Rhea Mons and Theia Mons. Ishtar Terra (Australia-size) and Aphrodite Terra (Africa-size) are the only large continents. (The International Astronomical Union has named almost all of the Venusian features after women, historical and mythological. The Alpha, Beta and Maxwell appellations remain from earlier Earth-based studies.)

Despite the failure of the IR radiometer on February 14, 1979, the atmosphere was seen to be composed of three broad regions: lower haze (31–47km), main cloud deck (47–70km) and upper haze (70–90km). The atmospheric probes went below the 31km lower limit of this instrument. Indirect evidence strongly suggests that the clouds are composed mainly of sulphuric acid. The temperature of the upper atmosphere varied by 250°C between the day and night sides, indicating strong winds blowing across the terminator. Wind speeds at the cloud tops were found to be 100m/sec at the equator. The fall in the observed quantity of sulphur dioxide in the atmosphere over the mission suggests that PVO arrived soon after a large volcanic eruption. Signals picked up on the 100Hz channel of the electric field detector indicate lightning, often associated with volcanism.

PVO confirmed that there is little, if any, magnetic field. The solar wind therefore interacts directly with the ionosphere, which at several hundred km is much thinner than Earth's. PVO will continue to map this region over a solar cycle, being healthy enough to survive until it enters the atmosphere of Venus and is destroyed in 1992. Controllers stopped using hydrazine on July 27, 1980 (Orbit 600, 4.5kg remaining), to hold the perigee steady; this height rose to 2,270km on June 30, 1986, and will now fall again until May 1992, when the remaining propellant will be used for about 90 days to delay atmospheric entry. The radar (switched off on March 19, 1981, after Orbit 834) will be usable again from spring 1991 as perigee drops below 700km, bringing new southern regions within reach. The instruments will be able to sample the atmosphere at low altitudes before the craft is destroyed.

PVO was used in February 1986 to make valuable ultra-violet observations of Halley's Comet as that body flew through perihelion and became unobservable from Earth. Future comets might also be detected. The gamma-ray burst detector will continue to provide another viewpoint on the mysterious bursts of high-intensity radiation, discovered in 1973, that sweep across the solar system from deep space.

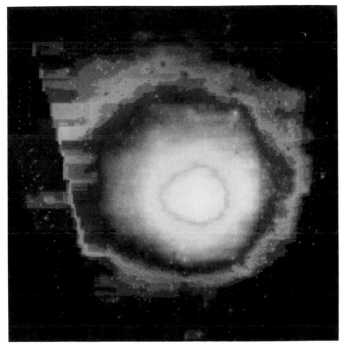

PVO was able to observe Halley's Comet while it was invisible from Earth. The UV spectrometer required 20,000 scans during February 2–5, 1986, to produce this image of the 12.5 million km hydrogen coma.

Pioneer Venus 2

Launched: 0733 GMT August 8, 1978
Vehicle: Atlas-Centaur 51 (Atlas No 5031D)
Site: ETR 36A
Spacecraft mass: 875kg at launch
Destination: Venus
Mission: Atmospheric probes
Arrival at target: December 9, 1978
Payload: Large Probe (316.5kg)
        3 Small Probes (93kg each)
        Bus:
           Neutral-mass spectrometer (6.3kg)
           Ion-mass spectrometer (3.0kg)
        Large Probe:
           Neutral-mass spectrometer (10.9kg)
           Gas chromatograph (6.30kg)
           Solar-flux radiometer (1.59kg)
           Infra-red radiometer (2.63kg)
           Cloud particle size spectrometer (4.35kg)

           Atmospheric structure experiment (2.31kg)
           Nephelometer (1.09kg)
        Small Probes, each:
           Atmospheric structure experiment (1.22kg)
           Nephelometer (1.09kg)
           Net-flux radiometer (1.09kg)
End of mission: December 9, 1978
Notes: First US Venus atmospheric probes

While Pioneer Venus 1 (PVO) was designed to return long-term remote sensing data on the planet, its companion spacecraft carried a set of four protected probes along a Type 1 path to sample the Venusian atmosphere directly on both sides of the day/night terminator. The 2.53m-diameter bus, similar to the orbiter but without a high-gain antenna and retromotor, itself housed instruments that would return data before its unprotected structure burned up in the higher atmosphere as it was rapidly decelerated from an entry speed of 42,000km/hr. Since the bus would plunge in directly, its data-handling system possessed no memory but returned data in real time at 8–2,048bits/sec, with 1,024bits/sec being used during entry. There were fore and aft-pointing omni antennae, as on the orbiter, plus an aft medium-gain antenna that would point Earthwards at entry. Power from 6.9m² of 2 × 2cm solar cells,

The PVO multiprobe craft (foreground) and orbiter (background).

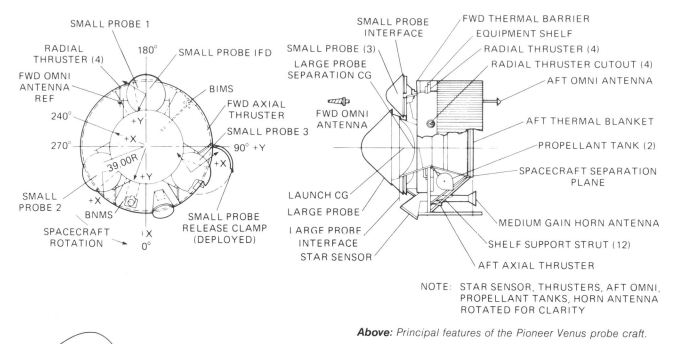

Above: *Principal features of the Pioneer Venus probe craft.*

Left: *Entry pattern for the five atmospheric probes.*

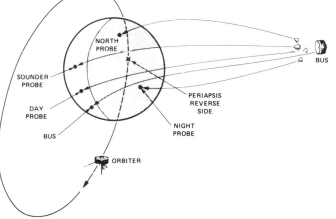

generating 241W at Venus, was also used during cruise to maintain a constant temperature inside the probes, which carried only batteries.

The Centaur left Pioneer Venus 2 with a Venus miss distance of 14,000km, but thruster burns on August 16 produced a ΔV of 2.25m/sec and brought the aim point back to the planet. Engineering telemetry from the four probes was routed via the bus until they were cut free to travel to their different destinations over the planetary disc. Until they began their entry sequences there was no contact with Earth. The Large Probe was the first to go – at 0237 GMT on November 16, while 11.1 million km from Venus – pushed out by springs from its central position towards entry just on the day side near the equator. None of the probes could be commanded from Earth once an internal timer was set and then started by release. Thereafter, each timer would issue its orders irrespective of whether the probe had reached the planet or not. The sequence was so vital to the success of the mission that three people had the job of independently calculating the timers' initial settings. The Large Probe was also the first NASA spacecraft to carry a microprocessor, an Intel 4004 in the neutral-mass spectrometer.

The three Small Probes were released simultaneously at 1306 GMT on November 20 while 9.3 million km from Venus, their restraining clamps firing open to allow the bus' 48rpm spin to spread them out like pellets from a shotgun. By the time they reached Venus, one would be on the night side at high northern latitude (North Probe), a second deeper into night but close to the equator (Night Probe), and the third just on the day side at mid-southern latitude near the bus entry point (Day Probe).

The Large Probe was the first to hit Venus. Its timer began a warm-up sequence at 1615 GMT on December 9 and at 182927 allowed the first burst of telemetry, which took three minutes to cross the gulf to Earth. This probe was about 1.5m in diameter, with a 73cm diameter pressure vessel protected by an aluminium heatshield covered in ablative carbon phenolic material. An aft cover of fibreglass honeycomb carried a Teflon centre section to allow data transmissions through from the pressure vessel. All of the experiments and the electronics were carried inside this vessel on two beryllium shelves kept at Earth atmosphere pressure in nitrogen, although a gas bottle allowed the pressure to be raised by 4.1N/cm² during the descent to help fight the increasing external pressure (90 Earth atmospheres close to the surface).

The Large Probe's interior was lined with a 2.5cm-thick layer of Kapton for further thermal control. The four remote-sensing instruments looked out through nine windows in the shell (eight of sapphire and one a 13.5 carat diamond for the IR instrument), while the mass spectrometer and gas chromatograph could sample the atmosphere directly through two inlets.

The entry began formally at 200km, deceleration peaking at 280g as air drag slowed the probe down from 42,000km/hr to 727km/hr in 38sec. During the 62sec radio blackout the 3,072-bit onboard data-storage capacity was used until a real-time link at 256bits/sec using the 2.3GHz S-band antenna could be established with Earth. Power since separation from the bus had come from a 40A-hr silver-zinc battery.

A small mortar fired on the side to drag out a pilot parachute which in turn pulled off the probe's aft cover. This tugged out the main 'chute from the aeroshell at 1815, and the heatshield was dropped to leave the pressure vessel and its scientific instruments operating under the deployed canopy 64km above the surface. Small vanes around the vessel's circumference maintained a slow rotation of less than 1rpm to scan the instruments around. After a fall of 19km in 16½min, the 'chute was cut free to allow the vessel to free-fall for another 39min. It struck the surface at 4.4°N/304.0° longitude and a speed of 32km/hr, impact being indicated by a sudden loss of data.

Below: *The Large Probe.*

Bottom: *Entry sequence for the Large Probe.*

Right: *External features of the Small Probes.*

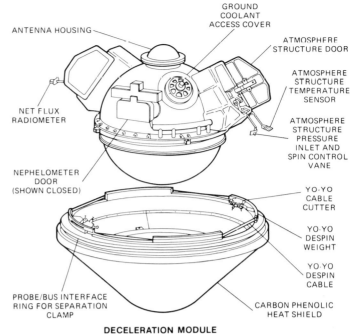

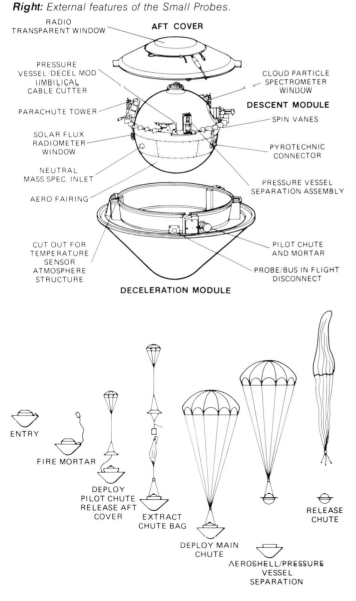

The simpler Small Probes followed only minutes later, spread over a six-minute period and undergoing peak decelerations of 200–565g. Unlike those of the Large Probe, the heatshields and aft covers remained attached throughout the descents and there were no parachutes: the dense atmosphere slowed the probes to just 35km/hr at landfall (North Probe 59.3°N/4.8° longitude; Day Probe 31.3°S/317.0°; Night Probe 28.7°S/56.7°). The three identical Small Probes were only 76cm in diameter overall and their two-piece titanium pressure vessels were filled with xenon at Earth atmospheric pressure to keep the internal temperature below 50°C.

Five minutes before entry, small weights were unwound on the end of 2.4m wires to slow spin from 48 to 17rpm before being jettisoned. Three doors on the afterbody opened at about 70km for the three instruments to begin operations. The nephelometer used a light-emitting diode to illuminate nearby cloud particles, recording the backscattered light to build up a picture of vertical cloud structure. The atmospheric structure experiment carried temperature, pressure and accelerometer sensors, while the net-flux radiometer measured the difference in intensity of radiation from above and below the probes as they descended.

Though the Small Probes were not designed to survive impact, the Night Probe continued transmission for 2sec, while the Day Probe survived for 67½min as its internal temperature rose to 126°C and the single 11A-hr silver-zinc battery became exhausted. Each probe carried a 10W transmitter, allowing a 64bits/sec data rate down to 30km and 16bits/sec in the thicker regions of the atmosphere. Sensors on all four probes suffered difficulties – the temperature probes failed, for example – probably because of electrical interaction with the atmosphere or deposition of sulphur from the clouds.

The bus hit the upper atmosphere at 2025 GMT over 37.9°S/290.9° longitude and sampled the higher regions – not investigated by the protected probes – before burning up after 63sec at a height of about 120km.

The probes showed that while the upper atmosphere indeed

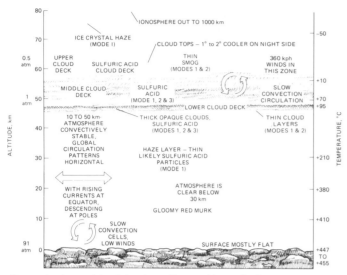

Atmospheric model arising from Pioneer Venus findings.

Diagram labels:
- IONOSPHERE OUT TO 1000 km
- ICE CRYSTAL HAZE (MODE I)
- CLOUD TOPS – 1° to 2° COOLER ON NIGHT SIDE
- UPPER CLOUD DECK
- SULFURIC ACID CLOUD DECK
- THIN SMOG (MODES 1 & 2)
- 360 kph WINDS IN THIS ZONE
- MIDDLE CLOUD DECK
- SULFURIC ACID (MODE 1, 2 & 3)
- SLOW CONVECTION CIRCULATION
- LOWER CLOUD DECK
- 10 TO 50 km ATMOSPHERE CONVECTIVELY STABLE, GLOBAL CIRCULATION PATTERNS HORIZONTAL
- THICK OPAQUE CLOUDS, SULFURIC ACID (MODES 1, 2 & 3)
- THIN CLOUD LAYERS (MODES 1 & 2)
- HAZE LAYER – THIN LIKELY SULFURIC ACID PARTICLES (MODE 1)
- WITH RISING CURRENTS AT EQUATOR, DESCENDING AT POLES
- ATMOSPHERE IS CLEAR BELOW 30 km
- GLOOMY RED MURK
- SLOW CONVECTION CELLS, LOW WINDS
- SURFACE MOSTLY FLAT

Altitude axis (km): 0, 10, 20, 30, 40, 50, 60, 70, 80
Pressure axis: 0.5 atm (~60), 1 atm (~50), 91 atm (0)
Temperature axis °C: –50, +10, +70, +95, +210, +380, +410, +447 TO +455

rotated once every four days, the 360km/hr winds dropped to almost zero at the surface. They revealed a fine haze layer (smog) at 70–90km and three cloud layers at 70–47km, the lowest of which was opaque. Between 10 and 50km there is little convection as we know it on Earth – Venus is almost stagnant – and following a haze layer at 30km the atmosphere is clear. (See Vegas 1 and 2 for an alternative Soviet analysis.) The clouds are composed largely of sulphuric acid droplets and the mass spectrometer inlet on the Large Probe became blocked with a droplet until the higher temperatures in the lower atmosphere boiled it off. It did allow measurement of the deuterium:hydrogen ratio, which turned out to be a hundred times higher than Earth's, suggesting that Venus had large water oceans in the past and possibly lost them following a steep temperature increase. Pioneer Venus confirmed that the greenhouse effect is helping to maintain high temperatures on the planet.

Some 96.5% of the atmosphere is carbon dioxide, with 3.5% nitrogen and less than 0.1% water. Sulphur dioxide accounts for some of the ultra-violet atmospheric circulation markings, but there must be another constituent as yet undetected. The Large Probe found the argon and neon levels to be several hundred times higher than expected, a result which will have to be accounted for in theories of solar system evolution.

ICE/ISEE-3

Launched: 151212 GMT August 12, 1978
Vehicle: Delta 2914 (Delta No 144)
Site: ETR 17B
Spacecraft mass: 478kg (422kg at comet encounter)
Destination: 1 Earth-Sun L1 libration point
2 Comet Giacobini-Zinner
Mission: 1 Halo orbit
2 Flyby
Arrival: 1 November 20, 1978, in halo orbit
2 September 11, 1985, closest approach 7,862km at 1102 GMT

Payload: Solar wind plasma
Plasma composition
Energetic protons
X-ray, low-energy electrons
Low-energy cosmic rays
Magnetometer
Plasma waves
Radio waves
+ gamma-ray burst and 4 cosmic-ray detectors irrelevant to comet encounter
End of mission: 2010+ (signal beyond range 1988–2008)
Notes: First comet encounter, first craft in halo orbit

International Sun-Earth Explorer 3 was launched into a halo orbit about the Sun-Earth L1 libration point 1.5 million km on the sunward side of Earth to provide information on the solar wind. Simultaneous monitoring by ISEEs 2 and 3 investigated the downstream effects of the Earth's magnetosphere.

ISEE-3, 1.7m in diameter, 1.6m high and spin-stabilised at 20rpm, generated 160W from solar cells mounted on its 16-sided body. Two 3m booms carry magnetometer and plasma-wave sensors, and four 49m antennae are used in radio and plasma-wave studies. There is no imaging capability or high-speed dust protection. ISEE completed its main mission in 1981 and Dr Robert Farquhar of NASA's Goddard Space Flight Centre proposed that it be shifted first into the Earth's magnetic tail and later manoeuvred towards a comet. The 75% of the original 91kg of hydrazine remaining was more than sufficient but the small 5W transmitter dictated a nearby target, Halley's Comet being too distant. On June 10, 1982, GSFC controllers began to use the twelve 18N thrusters to move ISEE-3 into the geotail. In October NASA approved a mission to Comet Giacobini-Zinner, a small, erratic body discovered in 1900, with a period of 6½ years. The cost of the first comet encounter was only $3 million.

ISEE/ICE during its 120km lunar flyby. NASA has donated the craft to the Smithsonian Institution for display following its possible recovery in 2014.

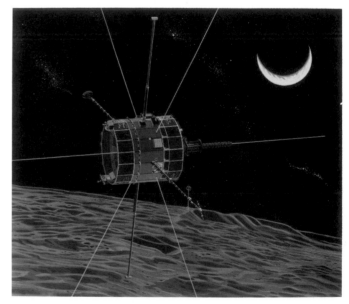

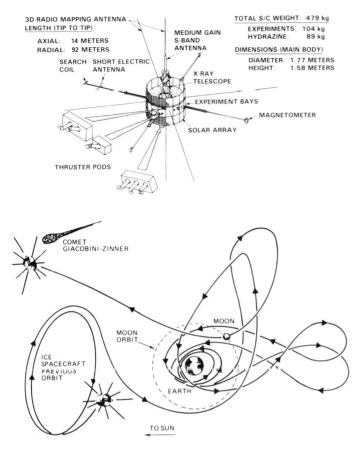

Top: ISEE-3 was designed to study fields and particles phenomena near the Earth.

Above: ISEE/ICE's complex series of manoeuvres to escape the Earth-Moon system.

tail, 25,000km wide and taking 20min to cross. Water and carbon monoxide ions were identified (confirming the "dirty snowball" nucleus theory), the plasma density increased 100 times over solar wind ambient, and the solar magnetic field was found wrapped around the nucleus like a hairpin. There were no dust detectors aboard but impacts were uncovered in the plasma wave data. The 20.7km/sec flyby was expected at least to damage the fragile antennae, but a rate of one impact per second showed the comet to be less dusty than models had predicted. At 1220 GMT, 250,000km from its first penetration, ICE crossed the bow shock on the outward leg.

ICE not only provided scientists with the first in situ cometary measurements, allowing revision of comet models before the Halley encounters, but it also flew 0.21 AU on the sunward side of Halley on March 28, 1986, to provide upstream solar wind data. Its trajectory will return it to the vicinity of the Earth in 2014, when it could be captured for analysis of its dusty coating.

Venera 11

Launched: September 9, 1978
Vehicle: Proton (D-1-e)
Site: Tyuratam
Spacecraft mass: 4,450kg at launch (1,600kg descent section)
Destination: Venus
Mission: Flyby/lander
Arrival: December 25, 1978 (landed 0324 at 14°S/299° longitude; bus flyby of 35,000km)
Payload: Bus:
      Plasma spectrometer
      Konus gamma-ray detector
      Sneg-2MZ gamma and X-ray burst detector (French)
      Ultra-violet spectrometer
      Magnetometer
      Solar wind detectors
      Cosmic ray detectors
    Capsule:
      Panoramic imaging system
      Sigma gas chromatograph (10kg)
      Mass spectrometer
      Gamma-ray spectrometer
      Groza lightning detector
      Temperature and pressure sensors
      Anemometer
      Nephelometer
      Optical spectrophotometer (4,500–12,000Å)
      X-ray fluorescence cloud aerosol analyser
End of mission: December 25, 1978, for lander. Bus?
Notes: Sixth Venus landing

A launch after the close of 1978's normal window prevented the Venera 11 bus from carrying sufficient propellant to enter Venus orbit: Clark estimates that there was only a quarter of the Venera 9/10 level. Flyby would however permit a longer period of data reception from the lander.

Venera 11 was launched first into a Type I trajectory but arrived at the target four days later than its Venera 12 twin. Course corrections on September 16 and December 17 estab-

Controllers ordered a series of lunar flybys to establish the necessary transfer trajectory. The first four produced swingbys on March 30 (19,570km), April 23 (21,137km), September 27 (22,790km) and October 21, 1983 (17,440km). The crucial fifth and final pass took ISEE-3 to within 120km above the Moon near the Apollo 11 site at 1845 GMT on December 22, 1983. The 28min eclipse period closed down the probe's systems (the single battery had failed in 1981) but ISEE-3 emerged unscathed and was renamed International Comet Explorer as it headed for a 200,000km encounter with the comet.

Some 2,000 pulses from the thrusters over 4½hr on June 5, 1985, shifted the trajectory to 26,550km behind the comet, where the particles and fields experiments could sample the tail (all of the Halley probes traversed in front of the nucleus). A second orbit correction was made on July 9, but tracking closer to the encounter showed that ICE would miss the tail axis by 800km. The thrusters were pulsed 299 times on September 8 to correct the error.

The first ions were detected a day and a million km out (further than expected), and at 0910 GMT on September 11 ICE crossed an ill-defined bow shock, whereas abrupt change in the magnetic field and plasma density had been expected. Further in, it found a turbulent region with interacting cometary and solar wind ions, and finally found the actual comet plasma

lished a path intersecting the day side, and once the descent module's batteries had been charged and its interior chilled the craft split into two on December 23. The bus executed a deflection burn to accomplish a 35,000km flyby as the entry sphere struck the atmosphere at 11.2km/sec. The new landers carried a wider range of instruments to analyse the atmosphere during descent. One of the major surprises was the discovery that the argon-36: argon-40 ratio was about 1, compared with about 1:300 on Earth. This single fact cast doubt on the theory that the primitive atmospheres of Earth, Venus and Mars formed from outgassing; did they result instead from the capture of gaseous material from the solar nebula? Each mass spectrometer took aboard 11 small samples from 24km down to the surface.

The gas chromatograph was a new instrument and originally its data were misinterpreted to suggest that there was no carbon monoxide in the atmosphere; a 1980 reanalysis revealed this gas for the first time. This same device found 0.5% water vapour at 44km and 0.1% at 24km during Venera 11's descent, whereas the Venera 12 version failed to detect any. An instrument error was strongly suggested when the optical spectrometer measured water absorption bands at 7,200, 8,200 and 9,500Å and found very little water: 0.02% at 50km and 0.002% at the surface. If all Venus' water vapour were suddenly deposited it would form a global layer only 1cm thick; evolution theories have to explain why there is so little.

The spectrophotometer showed the clouds to be relatively transparent – visibility several kilometres – with 3–6% of the sunlight reaching ground level. But since the light is scattered, an observer would not be able to see the Sun. Cloud aerosols collected on filters were analysed by X-ray fluorescence techniques and indicated that over 49–61km chlorine compounds were predominant, and not sulphur as expected (Veneras 13/14 found the reverse).

Data from the lander were recorded for 95min following touchdown, but it appears that both Venera 11 and 12 suffered severe instrument failures. Few surface results have been released and no images were produced. Photographs of the

landers showed TV units similar to those of Venera 9/10, and Soviet scientists have unofficially acknowledged that there were failures. Surface conditions were measured at 446°C and 88 atm for Venera 11. Both landers also recorded extensive thunder and lightning with the Groza ("thunderstorm" in Russian) instrument, a low-frequency (8–100kHz) spectrum analyser. Such events are often associated with volcanic activity, although the high cloudbase (about 50km) suggests that the lightning bolts would be cloud-to-cloud rather than to the ground.

It is not known for how long the flyby modules continued to return data from their instruments, but 27 bursts of gamma radiation and 120 solar flares were reported before the encounters by the Konus device, with its six scintillation counters. Results were correlated with the contemporary US Pioneer Venus findings.

See Veneras 9/10 for spacecraft description.

Left: This picture, released as a Venera 11/12 representation, clearly shows a camera mounting below the aerodisc.

Below: Radio bursts recorded by Venera 11 during descent may have been caused by lightning strokes.

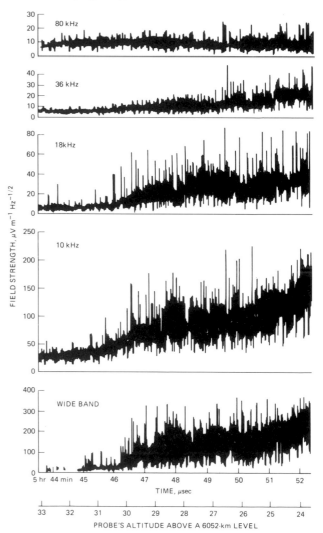

Venera 12

Launched: September 14, 1978
Vehicle: Proton (D-1-e)
Site: Tyuratam
Spacecraft mass: 4,461kg at launch (1,612kg descent section)
Destination: Venus
Mission: Flyby/lander
Arrival: December 21, 1978 (landed 0330 at 7°S/294° longitude; bus flyby at 35,000km)
Payload: As Venera 11
End of mission: December 21, 1978, for lander. Bus?
Notes: Fifth Venus landing

The second of the Soviet 1978 Venus craft also seems to have suffered partial instrument failure on landing. No surface picture has ever been released, whereas extensive results from data acquired during descent have been published (see Venera 11 for discussion of results).

Course corrections took place on September 21 and December 14: these second-generation Veneras fine-tuned the approach with a late correction about five days before capsule release, after which a deflection burn took the bus safely past the planet. The sphere entered the atmosphere at 11.2km/sec, and the parachute opened at 62km and was then jettisoned at 40km for the aerodisc to complete the ~1hr descent with the Sun 70° above the horizon. Some 110min of data were returned from a site about 800km north of the Venera 11 location in Phoebe Regio, including the 15min reverberation of a huge thundercap.

See Veneras 9/10, for spacecraft description, and Venera 11.

Venera 13

Launched: October 30, 1981
Vehicle: Proton (D-1-e)
Site: Tyuratam
Spacecraft mass: 4,363kg at launch, including 1,645kg descent section, 760kg lander
Destination: Venus
Mission: Flyby/lander
Arrival: March 1, 1982 (landed 7.5°S/303° longitude; bus flyby of 36,000km)
Payload: Bus:
    Magnetometer
    Cosmic ray detector
    Solar wind detectors
    Signe 2MS3 gamma-ray burst detector (French)
  Lander:
    Panoramic imaging system
    X-ray fluorescence spectrometer + drill
    X-ray fluorescence spectrometer for aerosols (62–45km)
    Pressure and temperature sensors (62–0km)
    Mass spectrometer, 9.5kg, 17W (26–0km)
    Gas chromatograph (58–0km)
    Groza 2 lightning detector (62–0km)
    Nephelometer (62–0km)
    Spectrophotometer (62–0km)

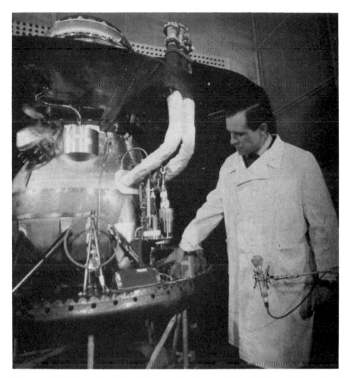

Venera 13/14 lander. The drill mechanism resides between the pair of cooling system pipes.

    Accelerometer (110–63km)
    Humidity sensor (50–45km)
    Prop soil mechanical/electrical probe
    Seismometer
End of mission: March 1, 1982, for lander; November 1982? for bus
Notes: Venus lander, first surface colour image, first soil analysis

The third pair of new-generation Venera landers carried a more extensive and improved set of instruments by comparison with their predecessors. The target sites were selected in consultation with US scientists using Pioneer Venus Orbiter radar maps. The original choices were shown by the US craft to lie in the same low plains to the east of the Phoebe Regio mountains. Dr Harold Masursky of the US Geological Survey visited the USSR in 1980 to suggest moving Venera 13 closer to Phoebe, where the rolling plains were possibly part of the ancient crust and were expected to be granite-like. Venera 14 would remain in the depressed lava-flooded area.

Two course corrections, on November 10, 1981, and February 21, 1982, allowed the chilled Venera 13 capsule to be separated during February 27 and the bus to complete an additional manoeuvre for a 36,000km flyby while acting as the data relay for the day-side lander. Capsule entry began at 0255 on March 1. Before the main parachute opened at 025631 only the accelerometers were active, registering atmospheric braking below 110km to help produce density profiles. Its useful data ended at 63km as the canopy deployed and other experiments became active (see payload list for heights).

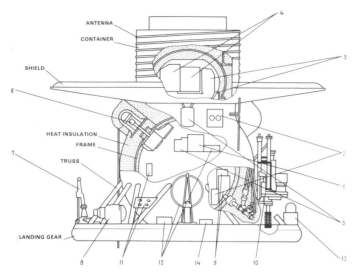

ANTENNA
CONTAINER
SHIELD
HEAT INSULATION
FRAME
TRUSS
LANDING GEAR

4
3
6
7
2
1
5
8 11 12 14 9 10 13
13

Above: *Venera 13/14 lander.* **Key:** *1 accelerometers 2 nephelometer 3 mass spectrometer 4 gas chromatograph 5 spectrometer and UV photometer 6 camera (2) 7 penetrometer 8 X-ray fluorescent spectrometer for aerosol analysis 9 X-ray fluorescence spectrometer for soil analysis 10 drill and soil delivery device 11 pressure and temperature instruments 12 radio spectrometer 13 hygrometer 14 experimental solar cells (measured surface illumination).*

Right: *Venera 13/14 penetrometer.*

Below: *Venera 13 returned the first colour pictures from the surface of Venus. Both show the ejected camera covers; the lower one the deployed penetrometer.*

The nephelometers on both landers, measuring aerosols from the backscattering of a light source, revealed a fine cloud structure, especially near the bottom deck, where 100m layers were detected. The improved spectrophotometer, with two spectral and one photometric channels covering 4,500–12,000Å, also operated all the way to the surface to measure the spectral and angular distribution of sunlight scattered by the atmosphere. One spectral channel observed above the module while the other scanned around. The 6,000 spectra obtained from the two missions showed that the clouds ended at 49km (Venera 13) and 47.5km (Venera 14), with three general layers above: a dense lower layer at 48–50km, a transparent mid-layer at 50–57km, and relatively dense clouds above 57km. Readings taken on the surface showed that 2.4% of incident sunlight was reaching ground level at the first site and 3.5% at the second.

The mass spectrometer, which took 7sec to analyse each atmospheric sample, was said to be up to 30 times more sensitive than its counterparts on Veneras 11/12. The gas chromatograph, with three columns instead of one, was significantly redesigned. An X-ray fluorescence spectrometer

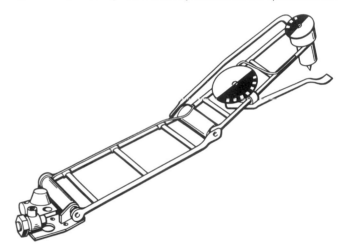

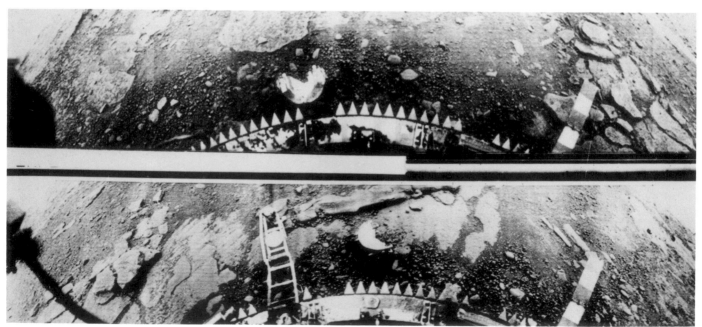

also analysed atmospheric aerosols and found sulphur to be the most abundant element, with chlorine next, the reverse of Venera 12's findings.

A humidity device using lithium chloride salt found a water content of 0.2 ± 0.04% over 50–46km, ten times more than registered by the mass spectrometer. The Groza electrical activity detector found that radio emissions below 10km were weaker (Veneras 11/12 had found none on the surface); scientists later concluded that the signals were possibly not from thunderstorms but from charging of the spacecraft itself during descent.

The parachute was released at 030526 and about 47km, and the landing, indicated by the accelerometers, followed at 035721. Transmissions continued for a further 127min, well beyond the 32min design life. Temperature and pressure were 465°C and 89.5 Earth atmospheres.

Both landers carried two cameras to provide 360° panoramas, and for the first time red, blue and green filters could be used to generate colour images. The pictures also had smaller pixels and a greater dynamic range than those of earlier landers. Each camera was housed 90cm above ground level, with a field of view of 37° × 180° centred 40° below the horizontal. Eight panoramas showed the lander to be in a field of orange-brown angular rocks and loose soil with evidence of possible chemical erosion. Successive images showed that soil was being blown on to the lander, indicating a surface wind speed of 1–2km/hr. The sky in each corner was orange, and the horizon appeared to be only about 100m away in both cases instead of the expected 1km. Scientists suggested that this effect was similar to a terrestrial mirage, and that as a result an astronaut would feel himself to be standing on a small body, although the phenomenon could be strongly height-dependent.

Also in the pictures were the ejected camera covers (12cm high, 20cm wide); a 9cm-wide colour scale; test patterns on the landing ring; protruding triangular "teeth," 5cm from tip to tip, that had helped to maintain stability during descent: and the deployed Prop probe, used to test mechanical and electrical surface properties.

A new device was the X-ray fluorescence spectrometer, designed to analyse a 1cc sample drilled from a depth of 3cm. This material was delivered to an internal chamber and pressurised to 1/20th Earth atmosphere and then subjected to radiation from plutonium-238 and iron-55. The induced radiation was detected by four proportional gas counters that found a soil similar to terrestrial leucitic basalt with high potassium content, rather a rare combination on Earth. A granite-like area similar to that found by Venera 8 had been expected. Venera 14's sample was similar to tholeiitic basalt with low potassium content, quite common on Earth.

The single-axis seismometer, able to detect vertical oscillations at a resolution of 26Hz, recorded no seismic events during Venera 13's active life on the surface, and only two small possible events were registered by its twin.

The Soviets acknowledged no further work by the flyby modules during Venera 11/12, but this time they announced that a large gamma-ray burst had been observed on May 11 (these mysterious energetic events from deep space were first detected by the US Vela nuclear-warning satellites in 1973). The main engines were fired again on June 10 and November 14, 1982, in an apparent demonstration of techniques for the Venera Venus/Halley's Comet missions of 1984–6.

See Veneras 9/10, for spacecraft description, Veneras 11/12 and Vegas 1/2.

Venera 14

Launched: November 4, 1981
Vehicle: Proton (D-1-e)
Site: Tyuratam
Spacecraft mass: 4,363.5kg at launch, including 1,645kg descent section, 760kg lander
Destination: Venus

Venera 14 surface images.

Mission: Flyby/lander
Arrival: March 5, 1982 (landed 13.25°S/310° longitude; bus flyby of 36,000km?)
Payload: As Venera 13
End of mission: March 5, 1982, for lander; November 1982? for bus
Notes: Eighth Venus lander

Venera 14 was the second successful mission of the lander/flyby pair launched during the 1981 window. But either its injection from parking orbit or the first course correction of November 14 was inaccurate, because on November 23 a second trajectory adjustment was carried out for the first and only time in the modern Venera series.

The usual manoeuvre just before reaching Venus came on February 25, 1982, in preparation for capsule release on March 3. Entry 1,000km from the Venera 13 site began at 0253 GMT, and the main parachute opened at 62–63km and 055400, activating the atmospheric instruments. The 'chute was released at 060253 and about 47km, and the ensuing free fall ended in surface contact at 070010 (indicated by accelerometer telemetry). There followed 57min of ground-level investigations. The higher pressure of 93.5 Earth atmospheres confirmed that this was a substantially lower location than Venera 13's, and the colour panoramas showed it to be on a 500m hill with small foothills in the distance. Surface temperature was 470°C and the site appeared harsher and more weathered than that of Venera 13.

See Venera 13 for spacecraft description and combined results.

Venera 15

Launched: 0238 GMT June 2, 1983
Vehicle: Proton (D-1-e)
Site: Tyuratam
Spacecraft mass: 5,250kg at launch
Destination: Venus
Mission: Orbiter
Arrival: October 10, 1983
Payload: Polyus V side-looking radar (300kg)
Omega radiometric system (25kg)
Fourier infra-red spectrometer (35kg)
Radio occultation experiment (10kg)
Cosmic ray detectors
Solar wind detector
End of mission: July 1984 (main mission)
Notes: Second Venus radar mapper

The 1978 Pioneer Venus Orbiter carried a small radar instrument capable of producing maps with resolution down to only 75km. The 1983 Venus window was used by the Soviets to fly two dedicated radar mappers built around modified Venera buses: the central core was lengthened by 1m (carrying about two tonnes of propellant, double that of the Venera 9/10 orbiters), and a 1.4 × 6m side-looking radar antenna working at 8cm wavelength replaced the usual descent sphere on top. Photographs show an increased number of gas bottles, probably to extend the life of the attitude-control system for lengthy

orbital operations, and two additional solar panels to satisfy the radar's power demands. The goal was to map Venus' northern hemisphere at 1–2km resolution in daily 150 × 7,000km strips 10° to the side of the orbital track.

Venera 15's course corrections were made on June 10 and October 1 in preparation for a braking burn beginning at 0305 on October 10 to establish an orbit of 1,000 × 65,000km, 24hr, 87° with perigee at around 60°N. The radar, as with Pioneer Venus Orbiter, would have operated only around closest approach each orbit, with the rest of the time devoted to battery recharging and transmitting the large quantity of data

Below: Venera 15/16, showing the extended central core and the radar antenna on top.

Bottom: North polar area of Venus as seen by Veneras 15 and 16. The circles are lines of latitude at 2° intervals.

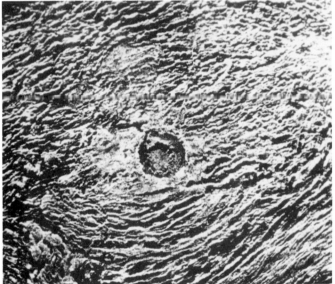

Top: Lakshmi Plateau, 3–5km higher than Sedna Plateau. The picture is centred at 65°/328°E and covers 1,360 × 1,960km. Note the mountains at right and the scan strips across the plateau.

Above: Crater in Patera Cleopatra. Picture is centred at 65°/78°E and covers 660 × 740km.

at 100kbits/sec via the main dish (enlarged by 1m for this mission) to the 70m Yevpatoria and 64m Medvezkye Ozyora antennae. Onboard processing allowed ground stations to produce quick-look images; the raw data were transmitted later for the full 8hr processing per strip to yield better resolution.

Mapping operations began during October 16 over the north pole but orbital considerations meant that the two craft had covered only from 30°N to the pole (115 million km²) by the time the main mission ended the following July. One orbiter (unidentified) was still operating in November 1984, but plans to shift its perigee to lower latitudes to extend the coverage were abandoned, perhaps because the attitude-control gas supply was insufficient.

The radar returns revealed impact craters, hills, ridges and other major surface features. There was disagreement over the interpretation of the results: Soviet scientists believed features around Maxwell Montes to be of impact origin, while their US colleagues saw them as volcanic. Images released by the end of 1985 show a lunar mare-like population of large (>50km diameter) craters. So dense is the atmosphere, the large craters must have been the work of impacting bodies at least 14km in diameter. Soviet scientists put the age of the rolling plains of Venus at a relatively young 1,000 million years. Beta Regio, the Soviets claim, has been confirmed as a rift system.

US scientists are less ready to draw broad conclusions from a 30% sample at limited resolution. The resolving power of the 300m Arecibo radio observatory in Puerto Rico is comparable with that achieved by the Soviet craft, and American investigators prefer to wait for NASA's Magellan radar mapper, due to fly in 1989 with four times better resolution, before trying to fully classify the planet's surface.

Apart from creating transmission and reception problems, the large volumes of data inherent in radar operations demand extensive processing. Two computers of the Soviet Institute of Radiotechnology and Electronics spent a year analysing 600km of tape, and an atlas based on the Venera observations is due for publication by "mid-1987". Twenty of the 27 maps had been completed by July 1986.

The Omega altimeter with its 1m antenna provided topographical coverage with a height resolution of 50m, and thermal data from the East German infra-red instrument plotted temperature variations over the northern hemisphere. Both instruments were mounted adjacent to the radar antenna. The average surface temperature was 500°C, with several hot spots of >700°C that could be sites of active volcanism. Clouds in the polar regions were found to be 5–8km lower than at the equator, and the atmosphere above 60km at 70–80° latitude is 5–20°C warmer than at 50° latitude. There was little diurnal variation at 50°.

See Venera 16.

Venera 16

Launched: 0233 GMT June 7, 1983
Vehicle: Proton (D-1-e)
Site: Tyuratam
Spacecraft mass: 5,300kg at launch
Destination: Venus
Mission: Orbiter
Arrival: October 14, 1983
Payload: As Venera 15
End of mission: July 1984 (main mission)
Notes: Third Venus radar mapper

Venera 16 was the companion spacecraft to the Venera 15 radar mapper. Course corrections on June 15 and October 5 allowed a braking burn to begin at 0622 on October 14 to establish a 24hr orbit, presumably with parameters similar to those of its predecessor. Mapping operations began on October 20.

See Venera 15 for spacecraft description and combined results.

Vega 1

Launched: 0916 GMT December 15, 1984
Vehicle: Proton (D-1-e)
Site: Tyuratam
Spacecraft mass: ~4,920kg at launch
Destination: Venus/Halley's Comet
Mission: Flyby/lander/atmosphere probe
Arrival: Venus: June 11, 1985, landed 7.2°N/177.8° longitude,
      bus closest approach 39,000km
    Halley's Comet: March 6, 1986, closest approach
      8,890km at 072006 at 79.2km/sec
Payload: Lander:
      Malachite mass spectrometer (64–48km)
      Sigma 3 gas chromatograph (64–40km)
      VM-4 hygrometer (64–48km)
      GS-15-SCV gamma-ray spectrometer (0km)
      UV spectrometer (64–0km; French)
      BDRP-AM25 X-ray fluorescence spectrometer
      + drill
      ISAV nephelometer/scatterometer (64–48km)
      Temperature and pressure sensors (130–0km)
      IFP aerosol analyser (64–48km)
    Balloon:
      Temperature and pressure sensors
      Vertical-wind anemometer
      Nephelometer
      Light Level/lighting detectors
    Bus:
      TV system (32kg)
      IR spectrometer (18kg)
      UV, visible, IR imaging spectrometer (14kg)
      Shield penetration detector (2kg)
      Dust detectors (9kg)
      Dust mass spectrometer (19kg)
      Neutral gas mass spectrometer (7kg)
      Plasma energy analyser (9kg)
      Energetic particle analyser (5kg)
      Magnetometer (4kg)
      Wave and plasma analysers (8kg)
End of mission: June 11, 1985 (lander)
               June 13, 1985 (balloon)
Notes: First Venus balloon, first close Halley flyby

Soviet scientists took advantage of the appearance of Halley's Comet by scheduling two Venera probes to continue on from Venus for the first close encounter. The two vehicles (combined mass 9,845kg) also each carried a 115kg balloon package for release into the planet's atmosphere, as well as typical landing capsules. The original intention had been to fly orbiters with large French balloons, but the mass/energy requirements proved too great and the atmospheric portion was deleted. Smaller Soviet balloons were then added when the mission was reoriented to allow the bus to encounter Halley.

The mission, named Vega from Venera-Gallei (the Russian language having no "H"), was announced well in advance, partly because of heavy international co-operation in the instrumentation and data analysis. The two launches also provided Western observers with their first clear views of the well tried Proton booster. Following the successful completion of both Halley flybys in March 1986, Roald Sagdeev, director of Moscow's Space Research Institute, revealed that both craft had sufficient propellant remaining to reach other comets or

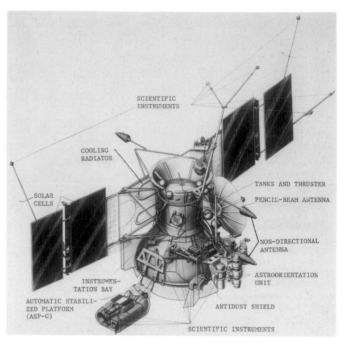

The Vega bus in flyby configuration.

asteroids, an option recognised in the planning stages but unpublicised until the main mission had ended.

The Vegas were based on the second-generation Venera vehicles introduced by Venera 9/10 in June 1975. They would be the last, the 1988 Mars probes being based on new designs. The three-axis-controlled cylindrical-conical bus, with an end diameter of 2.5m and height of 3m, carried two extra solar panels (as did Veneras 15/16) to bring the total area up to 10m², a 5m magnetometer boom, and 5m² of dust protection around the body and base consisting of a dual bumper shield (with a 0.4mm-thick front metallic layer) to ward off particles of up to 0.1g. Even so, being a modified Venus flyby module, it dared not venture closer than 6,000km to Halley. The final major addition was an 82kg scan platform near the base housing the twin TV system, IR spectrometer and three-channel spectrometer.

The 1,200mm-focal-length, f/6.5 and 150mm-focal-length, f/3 cameras used 512 × 512 and 512 × 256-pixel charge-coupled device (CCD) arrays simultaneously to produce 150m resolution at 10,000km in an effort to pinpoint the cometary nucleus for Giotto's closer visit. This pathfinder role for Vega was essential to the success of the European probe. Since Earth could not provide real-time control, the TV system included an 816kbit-memory computer to lock on to the expectedly bright nucleus and move the platform at up to 1°/sec. Data from the 14 bus instruments would be transmitted at 65,536bits/sec in real time at encounter, with 5Mbit-capacity tape recorders also saving the data for later replay through the 3,072bits/sec channel.

Vega 1 was slotted into a 168 × 205km, 51.6° parking orbit and dispatched into a Type II Venus trajectory by the Proton's fourth stage some 70min after launch.

The 1,500kg, 2.39m-diameter descent module separated on June 9, 1985, and hit the atmosphere 125km up in the early hours of June 11 as the bus flew on, having manoeuvred to

Above: *Vega lander and descent sphere displayed at the 1985 Paris Salon. The helium bottles and other balloon equipment can be seen mounted on top of the aerodisc, below which is an airflow stabiliser (carried for the first time). The tower-like drill mechanism is at front left of the pressure sphere, with the two shiny cylinders of the hygrometer at front. The large cylinder at rear right is the gas chromatograph; the pipe in front of it is the UV photometer. (Graham Turnill)*

Below: *Vega balloon and instrument package.*

miss the planet. On the way down, a small package separated at about 61km and opened a parachute at 55km. This was the first balloon to float in the clouds of Venus and radio back its findings to Earth. Prof Jacques Blamont of the French space agency had proposed the project in 1967, only to have the French-built balloons deleted in 1982.

The 3.54m-diameter Teflon-coated plastic balloon was inflated at 54km by 2kg of helium. Some 13m below it hung a slim, 1.2m long, 14cm-diameter 6.9kg gondola with a 4.5W transmitter operating at 1.667GHz. The balloon was radio-transparent, but even so signals faded down to 2W at times. Ballast was released at 50km and the balloon bobbed back up to its 54km operating altitude. This was in the middle and most active of the planet's three cloud layers, where scientists hoped to learn more of the atmospheric circulation and how it was driven. In operation the balloons massed 20.82kg (Vega 1) and 21.11kg (Vega 2), losing about 5% as gas leaked during the mission.

The three-section gondola was coated with a special white finish to ward off the corrosive atmosphere. The 30°-angle, 37cm-long upper cone acted as the antenna and the tether attach point. Suspended below on two straps was a 40.8 × 14.5 × 13cm mid-section carrying the transmitter, electronics and a deployed carbon-fibre arm with two temperature sensors and a four-bladed polypropylene anemometer propeller to measuring vertical wind speeds. The lower, 9 × 14.5 × 15cm, unit carried only the nepholometer, with its 9,300Å LED source for backscatter measurements of cloud particles. The 1kg of 250W-hr lithium batteries was expected to provide an active life of 46–52hr (collecting 22½hr of data). Information could be transmitted in 270sec bursts at 4bits/sec from the 1,024-bit memory every 30min at best, a frequency reduced later in the mission to conserve power.

Twenty antennae around the Earth, including six in the USSR (a 70m dish at Ussurisk in the far east was built specifically for Vega), formed a French-led network using very-long-baseline interferometry (VLBI) to measure the balloons' position to within 10km and their horizontal speed to within 3km/hr from a distance of 108 million km. Surprisingly, windspeeds in this 40°C, 0.5 Earth atmosphere region averaged 240km/hr instead of the expected 30–50km/hr.

Balloon 1 was released into the local midnight area just north of the equator and survived for 46½hr as it drifted 11,600km (11,100km for Balloon 2) over to the morning side. Both encountered numerous downward gusts averaging 1m/sec but reaching up to 3½m/sec, which regularly bounced them 200–300m vertically. Balloon 1 suffered mainly at the beginning and end of its life, whereas Balloon 2 initially found a quiescent atmosphere that grew more unsettled after 20hr. Beyond 33hr it encountered strong downdrafts for 8hr that at times pushed it 2½km down to the 0.9 Earth atmosphere level. This phase, when its horizontal speed was also affected, could be connected with passage over the highest areas of Aphrodite.

Balloon 1's light detector registered a regular increase in light levels after 33.8hr, some 3hr and 7° longitude before crossing the terminator, but neither craft conclusively picked up lightning flashes. No nephelometer data were returned by Balloon 2 and the other failed to find any clear regions in the cloud levels. The two sets of thermometers indicated a consistent 6.5°C temperature difference as they floated over more than 100° longitude, possibly because they remained in distinct and separate air masses. The cooler Balloon 2 data were

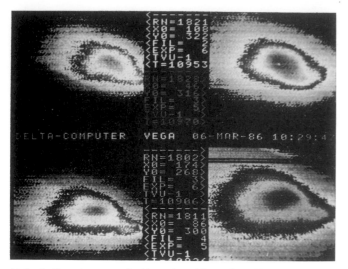

Halley's Comet as seen by Vega 1 on March 6, 1986.

confirmed by Lander 2 as it passed through on its way to the surface.

The balloons had been expected to burst on the bright day side but battery failure, indicated by fading signals, intervened first. By the time Balloon 1 expired, the lander had long ended its mission, transmitting 56min of data from its nighttime site in the Mermaid Plain north of Aphrodite. The smooth, low plain was typical of half the planet but no lander had ever analysed a sample from such a surface. Unfortunately, the drill began its sequence 10–15min before landing, while still 16km high, so no fragments of Venus were taken inside the capsule for examination by the X-ray spectrometer using iron-55 and plutonium-238.

Since there were no cameras aboard (Earth tracking of the balloon required a night-side descent), the failed soil sampler and the gamma-ray spectrometer were the main surface experiments. But the atmospheric descent had been put to good use. A French-Soviet mass spectrometer took in cloud samples and detected sulphur, chlorine and, possibly, phosphorus. It seems that there is some free sulphur, giving the clouds a yellow tinge. The gas chromatograph on both probes indicated 1mg of sulphuric acid in every cubic metre of air over 48–63km, a result corroborated by the mass spectrometer (which failed on the second lander).

The two experiments looking at light diffusion and the range and density of cloud particles produced fundamentally different results from the US Pioneer Venus probes. The clouds were likened to a thin terrestrial fog (albeit a highly corrosive one!) with most particles smaller than 1μm. There were two main cloud decks (not three, as suggested by Pioneer Venus), at 50km and 58km and some 3–5km thick. Nor did the clouds clear at 48–49km, persisting instead down to 35km. The two Vegas produced very similar results, so there could have been significant changes in the atmosphere since the December 1978 US findings.

The landers, like the balloons, also found substantial background IR radiation, and surface temperature and pressure for Vega 1 of 468°C and 95 Earth atmospheres. It was later admitted that the temperature value had been calculated since no direct reading was available.

Meanwhile, the Vega 1 bus had flown by Venus to begin its 708 million km voyage around the Sun to Halley's Comet. The final course correction was made on February 10, 1986, and two days later, 150 million km from the comet, the scan platform and its instruments were unlocked from their launch position against the side of the bus. On February 14 colour images of 807 million km-distant Jupiter were generated to calibrate the TV system.

Vega 1 began its studies proper of the comet at 0610 GMT on March 4, when "several dozen" images were obtained from 14 million km in a 1½hr session. The coma was little more than a dot but pictures such as these would help to pinpoint the nucleus for Giotto's encounter on March 13/14. A similar sequence was obtained a day later from 7 million km.

Vega's telemetry system switched over to high rate 2hr before closest approach. The probe plunged into the gas and dust-laden cloud around the coma itself, and in the 3hr encounter returned more than 500 images taken through different colour filters. It was battered by dust but no instruments were knocked out and the solar panels lost only part of their capacity. The Soviets later revealed that Vega 1's appeared to be slightly out of focus.

The images returned to the space institute in Moscow appeared to show two bright areas that were interpreted as a double nucleus. Giotto's subsequent experience shows that Vega was actually witnessing dust jets and not the nucleus at all. When Soviet scientists had had more time to digest the data they declared Halley's nucleus to be a dark (albedo <5%), 14km-long irregular body surrounded by dense layers of gas and dust. This was confirmed by subsequent processing. Images from both Soviet missions indicate a rotational period of 53 ± 3hr.

The dust detectors, picking up the first readings 320,000km out, indicated a dust production rate of 10^6 tonnes/day and there was an unexpectedly large number of small particles (10^{-8}m). The dust mass spectrometer detected three general compositions: similar to carbonaceous chondrite meteorites (carbon, oxygen, sodium, magnesium, silicon, calcium, iron); clathrate ice (water or carbon dioxide ice with other molecules trapped in its crystal lattice); and material with higher carbon and nitrogen concentrations.

The IR spectrometer's measurement of the nucleus temperature at 300–400K does not square with the 180–200K expected for an ice body. There is thus either a black porous layer about 1cm thick covering the surface, or a physically thin but optically thick layer of dust suspended just above it. The dark surface layer would explain the localised jet activity, dust particles being carried away with the sublimating water.

Water and carbon dioxide are the main parent molecules but there were signs of others, possibly organic. Overall, the observations suggest a similarity with some of the Saturnian and Jovian moons, supporting the theory that comets could have formed in this region of the solar system.

Subsequent imaging sessions were held on March 7 and 8 to provide data on the outward leg as Vega 1 sailed off back into deep space, having achieved the first close encounter with a comet. One Soviet craft could make distant surveys of asteroid Adonis in 1987.

Vega 2

Launched: 0914 GMT December 21, 1984
Vehicle: Proton (D-1-e)
Site: Tyuratam
Spacecraft mass: ~4,920kg at launch

Destination: Venus/Halley's Comet
Mission: Flyby/lander/atmosphere probe
Arrival: Venus: June 15, 1985, landed 6.45°S/181.08° longitude, bus closest approach 24,500km
Halley's Comet: March 9, 1986, closest approach 8,030km at 0720, speed 76.8km/sec
Payload: As Vega 1
End of mission: June 15, 1985 (lander); June 17 (balloon)
Notes: Second Halley's Comet close flyby

The Soviets adopted their usual philosophy of launching identical pairs of planetary probes to increase the chances of success. The Vega 2 balloon followed a similar path and yielded results similar (although its nephelometer failed) to those of its predecessor, floating around one-third of the planet and transmitting data for 46½hr. It found a temperature consistently 6.5°C lower. At one point, flying over a 5km mountain peak in Aphrodite, it hit a strong downdraft and plummeted 2½km.

The lander came down 1,500km SE of Vega 1, in the northern region of Aphrodite. Fortunately, it was in the midst of a patch well explored by Pioneer Venus Orbiter's radar which turned out to be 1½km above the mean surface and surprisingly smooth for the region. This indicates that the site is either very old (eroded) or very young (newly laid). The lander transmitted for 57min in the intense heat, announcing that the X-ray soil analysis had uncovered a rock type rare on Earth but found in the lunar highlands: anorthosite-troctolite, rich in aluminium and silicon but poor in iron and magnesium. The area appears to be the oldest tested on Venus thus far, and the high sulphur content (4–5% by mass) suggests chemical interaction with the atmosphere. Surface temperature and pressure were 463°C and 91 Earth atmospheres. The mass spectrometer failed to return any data during the descent.

Vega 2 image of Halley's Comet from a distance of 8030km, 2 sec before closest approach.

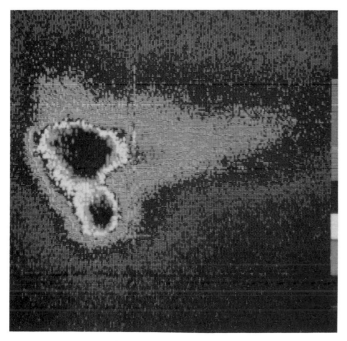

As with the balloon and lander missions, the flyby of Halley's Comet followed a pattern very similar to that of Vega 1. This time, however, the encounter distance was reduced slightly, increasing the risk of dust damage.

The approach trajectory was sufficiently accurate to obviate the planned final course correction on February 17, 1986, and on March 7 the first 100 images were relayed from a distance of 14 million km. The close approach in the early hours of March 9 yielded 700 pictures of greater clarity than those from Vega 1; there appears to have been less dust for this transit outside the coma and there was certainly less damage to the spacecraft. Subsequent image processing exposed the nucleus, even though a failure in the scan platform pointing procession shortly before encounter had brought back-up procedures into play, reducing the volume and quality of images.

Imaging sessions were also held on March 10 and 11, at 7 and 14 million km respectively, on the way out.

See Vega 1 entry for results.

Sakigake (MS-T5)

Launched: 1926 GMT January 7, 1985
Vehicle: Mu-3SII
Site: Kagoshima Space Centre (31.25°N/131.08°E)
Spacecraft mass: 130.1kg
Destination: Halley's Comet
Mission: Flyby
Arrival: March 11, 1986 (closest approach 6.99 million km at 0418 GMT
Payload: Plasma-wave probe (5.2kg)
Solar wind ion detector (2.0kg)
Magnetometer (5.4kg), on 2m boom
End of mission: 1990s?
Notes: First Japanese deep-space probe

The appearance of Halley's Comet provided an excellent opportunity for the Institute of Space and Astronautical Science to undertake Japan's first deep-space launchings. The 62-tonne all-solid-propellant Mu-3SII launcher was the most powerful at the investigators' disposal, but even so it allowed a mass budget of less than 150kg. This dictated the use of a simple, spin-stabilised probe carrying just a few kilograms of scientific experiments. There was no dust protection for a close approach.

A 64m radio dish was built for the mission at Usuda, 170km NW of Tokyo. Since Japan had no experience of controlling a deep-space probe, ISAS decided to fly a precursor mission, originally known as MS-T5 and then Sakigake ("Pioneer") after launch, to test operating procedures, the new launcher and the probe itself. Even though it would miss the comet by a wide margin, it could carry instruments to sample the solar wind and magnetic field for correlation with the comet's behaviour.

The two spacecraft were identical apart from the scientific payload. A carbon-fibre-reinforced plastic central thrust tube 51.5cm in diameter supported an aluminium honeycomb instrument/electronics platform on eight CFRP struts, with a 70cm-high, 140cm-diameter outer drum carrying 1,750 solar cells to provide up to 100W. A 2A-hr nickel-cadmium battery was also included.

The 32kN fourth-stage kick motor separated after placing the probe on a direct-ascent trajectory, leaving it spinning at 120rpm. Two blocks of 3N thrusters mounted on the top deck

Above left: Sakigake payload platform. On Suisei the black bar at rear right was replaced by the UV imager.

Above: Preparing Sakigake for launch.

and drawing on 10kg of hydrazine in two spherical titanium tanks were then used to slow the spin to 30rpm, and a 7.4kg, 2,000rpm momentum wheel provided fine tuning down to the cruise rate of 6.3rpm.

Sun and Canopus sensors provided attitude reference to hold the spin axis perpendicular to the ecliptic, except when trajectory adjustments were made using the thrusters (a total ΔV of 50m/sec was budgeted, using 5kg of hydrazine). A despun 80cm, 5W, S-band antenna relayed data at 64bits/sec at closest approach; there were also low and medium-gain antennae for command and control.

Sakigake carried no imaging device, but that aboard its twin required a spin rate of only 0.2rpm (the minimum allowable for adequate thermal control). While the 6.3rpm rate was suitable for solar wind measurements for 7–8hr/day, the momentum wheel had to slow the craft for 15hr of imaging per day, the data being stored in the 1Mbit bubble memory.

Sakigake was launched three days late because of ground equipment problems, at a cost of increasing the miss distance by about 3 million km. A miss of 7.6 million km on Halley's sunward side was produced by the launch, but course corrections three days later (ΔV 28.8m/sec) and on February 14, 1985 (ΔV 5m/sec), reduced it to 6.99 million km. Following engineering tests, all of the instruments were switched on before the end of February 1985.

Sakigake discovered that the solar wind was disturbed by the comet's presence at much greater distances than previously believed: 7 million km instead of 1 million, although twin

Suisei found the range to be 420,000km. Further analysis was necessary to unravel whether it was a real cometary effect or a coincidental change in solar wind conditions at time of closest approach. During Giotto's encounter Sakigake served as the reference spacecraft, allowing elimination of Earth atmospheric and ionospheric contributions to the variations in the European probe's radio signals from within the coma.

Both Japanese probes are expected to operate well into the 1990s. Earth re-encounter in 1992 will permit optional targeting to Mars, Venus or an asteroid/comet.

See Suisei.

Giotto

Launched: 112316 GMT July 2, 1985
Vehicle: Ariane 1 (V14)
Site: ELA 1, Kourou, French Guiana
Spacecraft mass: 960kg at launch, 573.7kg at flyby
Destination: Halley's Comet

Arrival: March 14, 1986, closest approach 605km at 000302 GMT; Earth encounter July 2, 1990
Payload: Halley Multicolour camera (13.5kg)
Neutral mass spectrometer (12.7kg)
Ion mass spectrometer (9.0kg)
Dust mass spectrometer; PIA (9.0kg)
Dust impact detector system (2.26kg)
Plasma analysis 1 (4.7kg)
Plasma analysis 2 (3.2kg)
Energetic-particle analyser (0.95kg)
Magnetometer (1.36kg)
Optical probe experiment (1.32kg)
End of mission: March 14, 1986 (main mission); craft in hibernation from April 2, 1986
Notes: First close cometary encounter; first European deep-space project

The honour of achieving the first close flyby of a comet passed to Europe's Giotto when the American effort was effectively killed on financial grounds. Numerous studies throughout the 1970s looked at comet flyby and rendezvous missions, often centred on Encke, Halley and Tempel 2, all examples of relatively bright, short-period comets. ESA accepted NASA's invitation to provide a small European probe to make a fast flyby of Halley in November 1985 before the main US section completed a rendezvous with Tempel 2 in July 1988. However, funding for the U3 ion thrusters was not forthcoming in NASA's FY1982 budget proposal of January 1980, leaving ESA's Science Programme Committee to decide the following March on a separate European ballistic craft.

The study for this mission was completed in May 1980, and the SPC decided that Giotto, in competition with five other

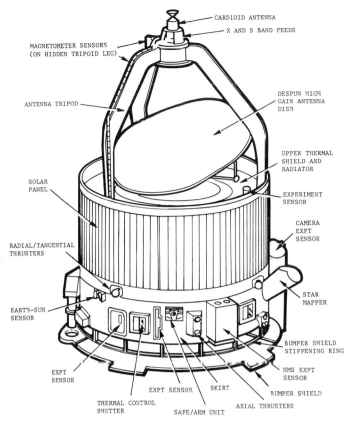

GIOTTO SPACECRAFT CONFIGURATION

Principal features of Giotto.

Giotto attached to its final stage and awaiting the fairing, at right background. The prominent black arm is the Plasma Impact Analyser. The camera turret is barely visible behind the dust shield at left.

projects, would be the next ESA scientific mission. Launch by Ariane 2 was only five years in the future, a remarkably short time to develop a complex spacecraft and mission plan. Experiment proposals had to be submitted by October 15, 1980, in time for a payload announcement in mid-January 1981. The probe was named after artist Giotto di Bondone, who used the appearance of Halley's Comet in 1301 as the model for the Star of Bethlehem in his 1304 *Adoration of the Magi* fresco in Padua's Scravegni chapel.

Giotto was based on the spin-stabilised Geos magnetospheric research satellites (Geos 1 was launched April 20, 1977, Geos 2 July 14, 1978), built by British Aerospace at Filton, Bristol. In July 1982 the company, as head of the STAR consortium, was named prime contractor for the £34 million spacecraft. No back-up was built, since any flight failure would leave insufficient time to prepare a second vehicle as Halley swept on through the inner solar system. The spacecraft was sized to match a Sylda 4400 geostationary transfer orbit launch aboard an Ariane 3, but the Giotto orbital constraints made it impossible for Arianespace to sell the upper payload space. The mission instead took over the Ariane 1 vehicle that had been originally destined for Exosat. It could have injected Giotto directly into heliocentric orbit without the large onboard kick motor (see later), but flight model testing had progressed so far that the removal would have severely disrupted schedules. Giotto thus retained the old launch sequence and motor.

Halley's Comet was chosen as Giotto's target because, of the 1,000+ known comets, it is unique in being young, active and with a well defined path – essential for an intercept mission. It also required a relatively low launch energy and would be easily observable from Earth during encounter. Its main drawback is its retrograde orbit (ecliptic inclination 162°), yielding a post-perihelion encounter speed of 68km/sec – equivalent to an Atlantic crossing in 1½min! At this speed a 0.1g particle could penetrate 8cm of solid aluminium, and Giotto would have had to carry a 600kg shield to stand a chance of survival. Hypervelocity tests and experience with Earth-orbiting satellites showed that a dual shield is far more efficient, however. Its first layer vaporises all but the largest particles, and the rear face can easily absorb the spreading cone of plasma and debris. Giotto was thus designed to travel into the coma behind a 1mm aluminium 6061-T4 alloy outer face and a 13.5mm-thick Kevlar sandwich 23cm apart and capable of stopping a 1g strike.

Spin-stabilised at 4rpm during encounter, Giotto was 1.867m in diameter across the shield and 2.848m high. It was built around a central aluminium thrust tube supporting three aluminium sandwich platforms: the upper carried the despun antenna and telecommunications equipment, the mid-platform housed four thruster propellant tanks and associated plumbing, while the lower level carried most of the experiments just behind the shield.

The 5,032 solar cells on the cylindrical body were to provide 190W at encounter, and four 16A-hr silver-cadmium batteries were included to meet peak demands and replace lost capacity if dust impacts degraded the array's output. The 1.47m-diameter, 20W S and X-band antenna at the top was canted at 44.3° to the spin axis to point at Earth during flyby. One drawback of using X-band (8.4GHz) – which permitted 40kbits/sec real-time data transmission – was the need for accurate pointing. An error of 1° would be enough to break lock with Earth, as indeed happened just before closest approach. The carbon-fibre-reinforced plastic tripod feeding the antenna also carried the magnetometer. Two low-gain antennae were used during near-Earth operations.

The duplicate sets of four radial and tangential 2N thrusters could provide a total ΔV of 170m/sec from the 69kg of hydrazine loaded, 60m/sec (19kg) of it to take out any aiming error after the Mage 1SB solid-propellant motor had fired. This burn would provide the impulse (1,400m/sec) for the eight-month journey. The motor was housed inside the thrust tube, firing through a central hole in the bumper shield which would then be closed by two quadrispherical aluminium shells. The two shields and shells weighed a total of 49.7kg.

Attitude reference depended on Earth and Sun sensors soon after launch, switching to the Sun sensor and a star mapper sensitive down to 2.4 magnitude further out (but which still used the bright Earth). Thermal control was maintained by surface finishes, thermal blankets, heaters and three thermal shutters on the outer surface.

The 10 experiments, weighing a total of 60kg, were:

- a camera for inner coma and nucleus imaging
- neutral, ion and dust mass spectrometers for composition analysis
- dust impact detectors to study dust output
- ion plasma analysers, electron analyser, energetic particle experiment, and magnetometer for plasma studies
- optical probe to study the dust and gas environment

The camera used a Ritchey-Chretien 998mm-focal-length

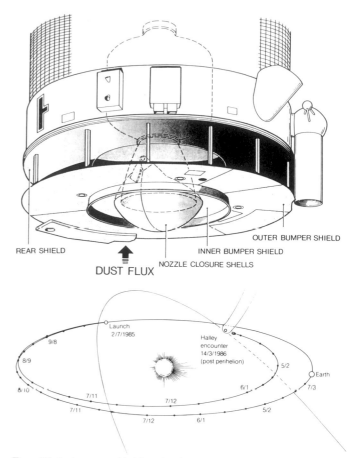

Top: Giotto bumper shield protection system.

Above: Giotto and comet trajectories.

f/7.68 telescope arrangement viewing, for impact protection, via a 45° mirror at the bottom of a rotatable turret. Two CCDs used several sensitive lines scanned across the comet image as Giotto rotated, the other lines in the arrays being used as efficient data stores. Some lines were covered with filters transmitting red and blue light, and one had a 56mm-diameter filter wheel with red, orange, blue and polarising filters. Expected resolution was 30m at 1,400km.

The dust detectors, provided by Britain's Kent University, made use of the bumper shield to assess the size of impacting particles. Three piezo-electric detectors – essentially microphones – were spaced 120° apart on the outer face to produce a 2m² sounding board. The shield was carefully calibrated so that the size of each impact could be determined (in reality, only a few were recorded by more than one detector). A similar device on the rear shield would pick up masses greater than 10^{-6}g after they had penetrated the first line of defence. Two other instruments on the front face provided coverage for smaller impacts. The Capacitor Impact Sensor (for masses less than 10^{-10}g) recorded the discharge of a 1,000cm² capacitor as it was perforated, and the Impact Plasma Detector (down to 10^{-17}g), covering 100cm², responded to the cloud of plasma generated by minute impacts.

The dust's elemental composition was determined by the

Impact Analyser (PIA), which used an electric field to sort the ions generated by dust striking a platinum target. The time taken for each type of ion to reach the detector depended on the atomic mass.

Giotto, spinning at 9.97rpm, was injected into a 190.5 × 36,000km, 7° geostationary transfer orbit. Twenty-two minutes after launch, authority passed to the European Space Operations Centre control room at Darmstadt in West Germany, which had only 32hr to assess the probe's status and prepare it for the final boost. Thruster firings increased the spin rate to 15rpm and then 90rpm before the French Mage solid motor with 374kg of propellant was ignited at 192317 GMT on July 3 for a 55sec burn as Giotto reached perigee after three revolutions. The impulse broke it free of Earth's grasp and established a heliocentric orbit that would miss Halley by 120,000km. The hydrazine tanks had been budgeted to provide a large correction because of the inherent inaccuracy of solid motors, but the path was so good that the first correction was cancelled.

Giotto was thoroughly checked out over the next four days, its spin reduced to 15rpm in two stages (consuming 1.62kg of propellant), the main antenna despun at 1002 on July 6, and the motor nozzle cover closed at 0730 on July 4 to complete the dust protection. Giotto's mass at the start of the cruise phase was 578.5kg, with 64.56kg of hydrazine remaining.

The camera was the first instrument to be switched on (August 10, 1985), with all experiments operating by October 13 in time to begin several dress rehearsals for the encounter. An Earth image taken during October 18 from a distance of 20 million km showed the camera to be performing to expectations, revealing cloud features over Australasia and Antarctica.

The first course correction was a 7.4m/sec pulsed burn on August 26 to refine the flyby to less than 4,000km. A 0.566m/sec correction was made on February 12, 1986, consuming 0.175kg of hydrazine over 7min 22.8sec.

TV observations by the two Soviet Vegas were essential for Giotto's close encounter. Earth-based observations produced a positional error of 1,500km, but data from the Vega 1 and 2 flybys reduced it to 180km and then 75km (99.7% level of probability). A final meeting of the principal investigators at Darmstadt on March 11 decided on a flyby of 500 ± 40km (68% probability), the error arising from uncertainty about the positions of both comet and probe. The distance was a compromise: the camera team required 1,000km ideally but no closer than 500km, while the others wanted the smallest separation possible, some even being unconcerned whether the probe survived or not. A 32min burn on March 12 reduced the previous 850 ± 70km miss distance to an actual flyby of 605 ± 8km (based on camera observations), the difference from the planned figure arising because the targeting assumed that the brightest region would be the nucleus.

The final, 8hr, dress rehearsal for the experimenters and controllers at Darmstadt was held during the night of March 10/11, with the instruments operating but Halley, at 20 million km, too distant for the camera. At 2100 on March 12 the first plasma analysis instrument detected the first Halley hydrogen ions 7.8 million km out. Images were also received that day, but then a bulldozer dug up a landline at Parkes in Australia (the main dish for the encounter) and contact was lost for a time.

At 1940 on March 13, 1,064,000km out, Giotto crossed the bow shock in the solar wind and the 4hr formal encounter began 35min later. Earth was 144 million km distant and data

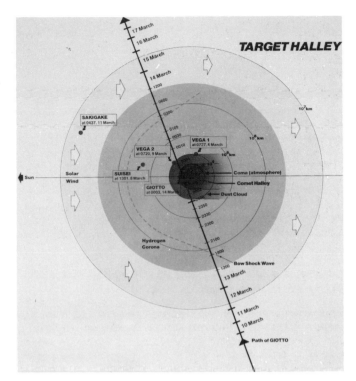

Giotto approach trajectory to the comet's necleus. Times and distances marked are pre-flyby expected values.

took 8min to cross the gulf. The camera had been turned on at 1943 and returned its first image at 2055. The dust investigators were becoming concerned at the absence of impacts when the first of about 12,000 struck 62min (290,000km out) before encounter. At 2358, with 20,100km to go, Giotto passed through the contact surface, where the solar wind is turned back by the cometary material, by which time the nucleus was so close that the camera had switched to tracking mode to follow the brightest object in its field of view.

The images being unveiled on the screen at Darmstadt were not readily interpretable because they showed only false-colour brightness levels; what appeared to be the core later emerged to be the base of a bright jet. Subsequent processing transformed the data into conventional colour images. The best of the 2,112 pictures, captured from 18,270km away, shows a lumpy, irregular body 15km long and 7–10km wide (6km had been expected), the full width being obscured by two large jets of dust and gas erupting into space on the sunward side. The dark side, with an unexpectedly low albedo of 2–4%, was quiescent, but enhancement revealed circular structures, valleys and hills over the entire surface. The jets seemed to originate in breaks in a dark crust that insulates the underlying ice from solar radiation. This picture of cometary surfaces strengthens the suspected link between exhausted comets and asteroids.

Images continued to within 1,372km, 18sec before closest approach. The rate of dust impacts had been lower than expected, but in the last few minutes it rose sharply and in the final few seconds there were 230 strikes as Giotto apparently penetrated one of the jets. Three of the front detectors peaked 26–28sec before closest approach; only 11 strikes reached

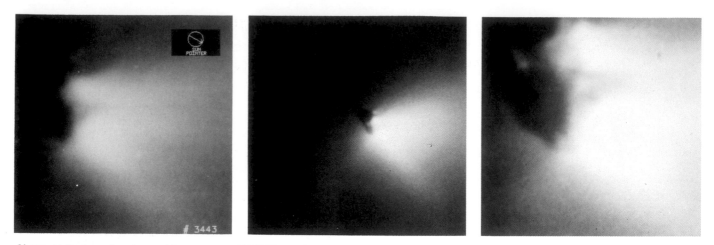

Above: *Halley's nucleus imaged from a range of 18,270km. In this, the best full image acquired, the structure is evident at top left, while the sunward side is obscured by two major jets. Subsequent processing has produced clearer images.*

Above centre: *Image 3418, taken from a distance of 25,700km and released in this form in October 1986, highlights the dust jets from the Halley nucleus.*

Above right: *Image 3475, produced at a distance of 9,500km. The 400m resolution is beginning to show crater-like depressions.*

the rear shield throughout, the first at −208sec. Just 14sec before closest approach Giotto was hit by a large particle (only 33mg striking the edge was sufficient to break Earth lock), which shifted the spin axis by 0.9°, beginning a 0.9°-amplitude nutating motion with a period of 16sec. The probe's spin period dropped from 4.010sec to 3.998sec (measured following recovery), and a permanent axis shift indicated that 600–700g of the structure had gone.

The screen at Darmstadt flickered and Halley was seen no more as Giotto's beam of data swung away from Earth 7sec before closest approach. Less than two minutes later, as Giotto sped away from Halley, intermittent signals were picked up (some useful data were later recovered). But it was a further 32min before the craft's nutation damper had killed the movement sufficiently for command lock to be regained. The camera had jammed at an angle of 45° (it was freed at the second attempt) and no more pictures were returned. It is not known if the mirror was destroyed, and further tests will be conducted before the decision is taken on an extended mission for the 1990s. Certainly, the star mapper's baffle was peppered with holes − sunlight could be seen streaming through it and it can be used only on the shadowed side − and the temperature rose 10–20°C in some sections because of thermal subsystem damage. The solar array, largely in the shield's shadow, lost only 5W of its 196W output. The energetic-particle analyser, optical probe and magnetometer continued to function, the impact plasma detector portion of the dust impact detector was not usable, the neutral mass spectrometer was lost, and one detector of the first plasma analysis system stopped working 1½hr after encounter. Some instruments had suffered high-voltage problems that indicated short-circuiting caused by plasma surrounding Giotto. Doppler and ranging measurements as part of the radio science experiment showed that the probe had been slowed by 23.2cm/sec over a 100sec period, reducing frequency by 4.7Hz and indicating a total impacting mass of 0.1–1g.

Water, as expected, was found to account for 80% by volume of all the molecules thrown out, equivalent to 10 tonnes/sec. Dust measurements indicated 3 tonnes/sec being ejected in seven identified jets; the largest grain detected, 44sec before closest approach, massed 40mg. The 10^{-14}g particles and smaller were much more abundant than anticipated (agreeing with Vega), but close in larger material was dominant. The final dust impact was picked up by the piezoelectric sensors 49min 13sec into the outward journey. The Plasma Impact Analyser showed that most dust was rich in H, C, N and O, and the ion mass spectrometer found an abundance of C^+ ions, some possibly arising from carbon atoms released directly from the surface. The suggestion is that Halley is covered in a layer of organic material, although the instruments were not capable of detecting complex molecules.

At 0200 GMT on March 15 Giotto had successfully completed its work and the experiments were turned off. The decision was made to expend some of the remaining 60kg of propellant in three manoeuvres to bring it to within 22,000km of Earth on July 2, 1990: March 19, 4hr 30min burn, 63.6m/sec, 19kg propellant consumed; March 20, 4hr 10min, 46.3m/sec, 14kg; March 21, 16min, 2.6m/sec, 0.8kg. An encounter with Comet Grigg-Skjellerup is now possible on July 14, 1992, as are meetings with other comets and asteroids. It is competing with other science projects within ESA for selection as a "new-start" in late 1988 or early 1989. But first it must be reactivated for an assessment of its instruments, particularly the camera. The probe was placed in hibernation on April 2, 1986, in a 0.7327 × 1.0400 AU orbit with a period of 305 days and ecliptic inclination of 23.91°.

Suisei (Planet-A)

Launched: 2333 GMT August 18, 1985
Vehicle: Mu-3SII
Site: Kagoshima Space Centre
Spacecraft mass: 139.5kg
Destination: Halley's Comet
Mission: Flyby
Arrival: March 8, 1986 (closest approach 151,000km at 1306 GMT)

Payload: UV imaging system (7.5kg)
Solar wind experiment (4.7kg)
End of mission: 1990s?
Notes: Second Japanese deep-space probe

The second Mu-3SII launcher performed so well that a trajectory adjustment planned for August 22 was cancelled, leaving Suisei ("Comet") with a miss distance of 210,000km. A 12m/sec correction on November 14, 1985, reduced the miss to 151,000km on the sunward side.

The main payload was an imaging system working in ultraviolet, where it would be able to study the vast hydrogen corona surrounding the comet and fluorescing under solar radiation. Beginning observations in November 1985, well before comet perihelion on February 9, 1986, it could watch the cloud grow, helping to determine the total hydrogen production rate resulting from the break-up of water vapour (determined from the mission to be 60 tonnes/sec at most).

Up to six images a day could be generated over a 15hr period, then read out from the 1Mbit bubble memory over 4hr at 64bits/sec. The pictures were built up as Suisei spun at 0.2rpm, using a 100mm-focal-length telescope providing a field of view of 2.5° for the 122 × 153 pixels of the charge-coupled device. The remaining 7hr each day could be devoted to measuring the distribution of 30–16,000eV electrons and ions as they streamed in from the Sun to interact with

Right: Suisei, with the cylindrical UV instrument at left.

Below: Launch of Suisei.

the cometary ionosphere. This was especially important around perihelion, when the Earth's view was obscured by the Sun. At closest approach Suisei would actually be inside the hydrogen corona.

The UV instrument was activated in early September 1985 and generated test images of Earth and O and B-type stars (strong UV emitters). It also produced faint images of Comet Giacobini-Zinner shortly after the September 11 ICE encounter. Routine imaging of Halley began in mid-November, when the hydrogen corona became visible, revealing brightness variations with a period of 2.2 ± 0.1 days that were interpreted as resulting from spinning of the nucleus. Halley moved too close to the Sun on January 10, 1986, and observations were resumed on February 9 around the comet's perihelion. Two days before closest approach, within the cloud, the camera was switched to photometry mode for brightness measurements looking towards the nucleus. At 1232 GMT on March 8, 34min before closest approach, the UV instrument was switched off to allow charged-particle measurements (the two systems could not operate simultaneously). Observations had been made since September 27, 1985, and the new sequence covering the outward leg showed that the solar wind was being slowed down to 70km/sec from 350km/sec in a region about 420,000km out (contrasting with Sakigake's 7 million km). Close in, the instrument picked up cometary water, carbon monoxide and carbon dioxide ions.

Continuous photometry measurements over 58hr from 0300 on March 21 showed there were at least two major and four minor outbursts on Halley, their 53hr period arising from the rotation of the nucleus. Their calculated positions tied in with the fact that the Soviet Vega 1 suffered greater dust damage than its later twin. In fact, attitude disturbances at 1254 and 1326 on March 8 at the time of closest approach indicate that Suisei itself suffered two dust impacts totalling several milligrams. The spin axis was shifted by 0.75° and spin period reduced by about 0.3%.

Both Japanese probes are expected to continue operating well into the 1990s, although the channeltron detector of Suisei's solar wind instrument will have degraded seriously by 1988/9. Earth re-encounter in 1992 will permit possible targeting to Venus, Mars or an asteroid/comet.

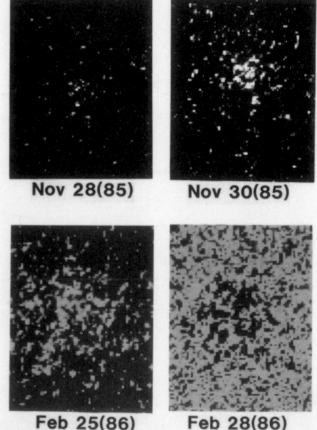

Nov 28(85) **Nov 30(85)**

Feb 25(86) **Feb 28(86)**

Top: Suisei's encounter configuration.

Bottom: UV images of Halley's hydrogen coma. The fourth frame was recorded at a range of 80 million km.

Appendix 1: future missions

The Soviets have recently been remarkably open about their plans for future deep-space missions, revealing that Mars rather than Venus will be the focus of attention. France will be a significant contributor to this programme, and a commitment to a Mars sample-return mission could provide the basis for USSR/US/ESA co-operation. The US programme has suffered uncertainties following cancellation of the Shuttle-Centaur combination, forcing a radical rethinking of America's deep-space missions, some of which have spacecraft already built. Listed below are the probes awaiting launch, missions with firm commitments, and some future studies.

Moon

The USSR has approved a 1991 flight of a lunar orbiter with up to 300kg of instruments for geochemical mapping of the entire globe. JPL's Lunar Geochemical Observer is the favourite for selection as the second Planetary Observer, a series of inner-solar system probes based on existing Earth-orbiting satellite designs (see Mars Observer). The other two candidates are the Mars Aeronomy Observer and Near-Earth Asteroid Rendezvous. Japan is studying a 1994 lunar polar orbiter in competition with a Venus orbiter: it would fire two or three penetrators into the far side to relay seismic data.

Mercury

Although a Mercury mission is not likely to be approved for some years, JPL has studied launch opportunities over 1990–2010. A combination of Venus and Mercury swingbys would be used to establish a 300km orbiter and possibly a subsatellite and surface landers. Windows will be available, in order of decreasing quality, in July 1994, July 1991, July 1996, August 2005, July 2004, July 1999, July 2007 and September 2002.

Venus

NASA's Magellan (formerly Venus Radar Mapper), originally scheduled for Shuttle Centaur launch in April 1988 to radar-map Venus at high resolution, has now been postponed to a late-April 1989 Shuttle-IUS launch. Japan's ISAS is studying a Venus-orbiting particles and fields craft for 1994 as the next Japanese space science project.

Mars

1 Two Soviet Phobos probes will be launched in a window that opens on July 15, 1988, for arrival at Mars in January 1989. They will carry out global geochemical mapping with a sophisticated array of instruments and techniques, including a 50m hover above moon Phobos to permit firing of the Lima D laser and Dion ion-beam experiments at the surface for analysis of the regolith by mass spectrometer. Two small surface landers could use X-ray fluorescence techniques for drill sample analysis. If the first probe is successful, the second could approach Deimos. First use of new-generation planetary craft.
2 Mars Observer, NASA's new start of FY1985 for August 20, 1990, launch by Shuttle or Titan with Transfer Orbital Stage into Sun-synchronous, 360km orbit, although postponement to 1992 to defer contractor payments was being considered in late 1986. It is planned to carry out a full Martian year (687 Earth days) of geochemical and mineralogical mapping and studies of atmospheric dynamics, especially moisture content, over a full season. First of NASA's Planetary Observer series; based on RCA's Satcom K comsat, with avionics from RCA's Tiros/DMSP metsat.
3 Vesta, Soviet probe for launch in November 1994 to deploy balloons and penetrators at Mars, with European module continuing on to asteroid/comet flyby. See comet/asteroid entry.
4 In June 1986 the Soviets announced that they were defining a 1996–98 Mars sample-return mission (MSRM), preceded by a 1994–96 lander/rover, although these dates now appear too early. JPL has also studied a November 1996 MSRM launch to return a 5kg sample in 1999, and is examining a Mars rover as the US contribution to a joint MSRM mission with the USSR.

Jupiter

The $800 million Galileo Jupiter orbiter/atmospheric probe should have been launched by Shuttle Orbiter *Atlantis* with a Centaur on May 20, 1986, for flyby of asteroid 29 Amphitrite on December 6, 1986, and arrival at Jupiter in late 1988. Galileo and Ulysses (see below) will use Shuttle-IUS combinations in the November 1989 and October 1990 slots. The decision on which will fly first was due in February 1987.

ESA's Ulysses solar probe, which will use a Jupiter encounter to fly high above the ecliptic plane and the Sun's poles for the first time, has been placed in storage at Dornier in West Germany. It should have been launched on May 15, 1986, by Shuttle Orbiter *Challenger* and Centaur in the same window as Galileo.

Saturn

Cassini, a joint ESA/NASA study for a Saturn orbiter/Titan atmospheric probe in 1990s. Second Mariner Mk II mission (see CRAF, below) could include asteroid flybys.

Comets/asteroids

NASA's Comet Rendezvous Asteroid Flyby (CRAF), proposed as an FY1989 new start project, would rendezvous with the short-period comet Tempel 2 in October 1996–December 1999 and fire instrumented penetrator into the nucleus. This is the first of the proposed Mariner Mk II missions. Flyby of asteroid 46 Hestia following launch in February 1993. Launch proposed September 1992. The Soviet Vesta probes could target several comets and asteroids following 1994 launch. Two of the Halley probes could see further service: a Vega could make a distant flyby of asteroid Adonis in 1987, while Giotto could encounter Comet Grigg-Skjellerup in 1992. ESA and NASA are very interested in returning a sample from a comet nucleus in about 2007 following launch at the turn of the century. Studies are under way.

Interstellar space

JPL is studying a 5,000kg "Thousand Astronomical Unit" (TAU) probe to penetrate up to 1,000 AU into interstellar space over a 50-year mission duration. Ion propulsion would accelerate it to 100km/sec by the time it had travelled 12,000 million km.

Appendix 2: solar system statistics

Sun and planets

	Sun	Mercury	Venus	Earth	Mars	Jupiter	Saturn	Uranus	Neptune	Pluto
Discovered	—	—	—	—	—	—	—	1781	1846	1930
Equatorial diameter (km)	1.39×10^6	4,878	12,104	12,756	6,787	142,800	120,000	50,900	48,600	2,200
Mass (Earth = 1)a	3.33×10^5	0.06	0.82	1.00	0.11	317.8	95.2	14.5	17.2	0.002?
Equatorial escape velocity (km/sec)	618	4.4	10.4	11.2c	5.1	59.5	35.6	21.3	23.8	1.1?
Equatorial rotation (days)b	25.4	58.65	243r	0.997	1.026	0.408	0.425	0.679	0.758	6.3
Density (water = 1)	1.41	5.43	5.24	5.52	3.54	1.32	0.70	1.27	1.77	1.6–2.6?
Perihelion (AU)	—	0.31	0.72	0.98	1.38	4.95	9.01	18.28	29.80	29.6
Aphelion (AU)	—	0.47	0.73	1.02	1.67	5.45	10.07	20.09	30.32	49.3
Sidereal period (Earth year = 1 = 365.256 days)	—	0.241	0.615	1	1.881	11.862	29.457	84.009	164.790	247.676
Orbit ecliptic inclination (°)	—	7.00	3.38	0.00	1.83	1.30	2.48	0.77	1.77	17.20

Notes: a Earth mass = 5.975×10^{24}kg. b Taking 1 day = 24hr. c For Moon = 2.38km/sec. r Indicates retrograde motion.

Satellites and rings

	Orbit radius (10^3km)	Maximum moon diameter (km)	Sidereal orbit period (days)	Inclination to equator (°)	Year of discovery
Mercury					
None					
Venus					
None					
Earth					
Moon	384.4	3,476	27.32	23.4	—
Mars					
Phobos	9.38	27	0.32	1.0	1877
Deimos	23.46	15	1.26	1.8	1877
Jupiter					
Ring	71.9–130	—	—	0	1979
Metis	128	40	0.295	0	1979
Adrastea	129	24	0.298	0	1979
Amalthea	181.3	270	0.498	0.45	1892
Thebe	221.9	110	0.675	0.9	1979
Io	421.6	3,630	1.769	0.04	1610
Europa	670.9	3,138	3.551	0.47	1610
Ganymede	1070	5,262	7.155	0.21	1610
Callisto	1880	4,800	16.689	0.51	1610
Leda	11,094	10	238.7	26.1	1974
Himalia	11,480	180	250.6	27.6	1904
Lysithea	11,720	20	259.2	29.0	1938
Elara	11,737	80	259.7	24.8	1904
Ananke	21,200	20	631	147	1951
Carme	22,600	30	692	164	1938
Pasiphae	23,500	40	735	145	1908
Sinope	23,700	30	758	153	1914
Saturn					
D-ring	67–74.4	—	—	0	1980
C-ring	73–92	—	—	0	1850
B-ring	92–117.4	—	—	0	1659
A-ring	122–136.6	—	—	0	1659
Atlas	137.7	40	0.61		1980

	Orbit radius (10³km)	Maximum moon diameter (km)	Sidereal orbit period (days)	Inclination to equator (°)	Year of discovery
Prometheus	139.35	140	0.61	0	1980
F-ring	140.3	—	—	0	1979
Pandora	141.7	110	0.63	0	1980
Epimetheus	151.4	140	0.69	0	1966
Janus	151.5	220	0.69	0	1966
G-ring	170	—	—	0	1980
Mimas	185.5	392	0.94	1.5	1789
1981S12	185.5	10	0.94	1.5	1981
Enceladus	238.0	510	1.37	0	1789
E-ring	180–480	—	—	0	1966
Tethys	294.7	1,060	1.89	1.1	1684
Telesto	294.7	34	1.9		1980
Calypso	294.7	34	1.9		1980
1981S6	294.7	20	1.9		1982
1981S10	350	15	2.4		1982
Dione	377.4	1,120	2.74	0.0	1684
Helene	377.4	36	2.74		1980
1981S7	377.5	20		0.3	1982
1981S9	470	20			1982
Rhea	527.1	1,530	4.52	0.3	1672
Titan	1,221.9	5,150	15.95	0.3	1655
Hyperion	1,481.0	410	21.28	0.4	1848
Iapetus	3,560.8	1,460	79.33	14.7	1671
Phoebe	12,954	220	550.34	150	1898
Uranus (ring widths given)					
Ring 1986U2R	37–39.5	—	—		1986
6-ring	41.85	1–3	—		1977
5-ring	42.24	2–3			1977
4-ring	42.58	2–3			1977
α-ring	44.73	7–12			1977
β-ring	45.67	7–12			1977
η-ring	47.18	0–2			1977
γ-ring	47.63	1–4			1977
δ-ring	48.31	3–9			1977
1986U7	49.7	40	0.33		1986
Ring 1986U1R	50.04	1–2			1986
ε-ring	51.16	22–93			1977
1986U8	53.8	50	0.38		1986
1986U9	59.2	50	0.43		1986
1986U3	61.8	60	0.46		1986
1986U6	62.7	60	0.48		1986
1986U2	64.6	80	0.49		1986
1986U1	66.1	80	0.51		1986
1986U4	69.9	60	0.56		1986
1986U5	75.3	60	0.62		1986
1985U1	86.0	170	0.76		1985
Miranda	129.9	484	1.41	4	1948
Ariel	190.9	1,160	2.52	0.3	1851
Umbriel	266.0	1,190	4.15	0.4	1851
Titania	436.3	1,610	8.70	0.1	1787
Oberon	583.4	1,550	13.46	0.1	1787
Neptune					
Triton	355	3,800	5.88	160	1846
Nereid	5,510	300	360.2	28	1949
Possible discontinuous ring					
Pluto					
Charon	19.4	1,160	6.38	?	1978

The existence of the numbered satellites is not certain and some might be deleted from future lists.

Index